Studies in Computational Intelligence

Volume 814

The series “Studies in Computational Intelligence” (SCI) publishes new developments and advances in the various areas of computational intelligence—quickly and with a high quality. The intent is to cover the theory, applications, and design methods of computational intelligence, as embedded in the fields of engineering, computer science, physics and life sciences, as well as the methodologies behind them. The series contains monographs, lecture notes and edited volumes in computational intelligence spanning the areas of neural networks, connectionist systems, genetic algorithms, evolutionary computation, artificial intelligence, cellular automata, self-organizing systems, soft computing, fuzzy systems, and hybrid intelligent systems. Of particular value to both the contributors and the readership are the short publication timeframe and the world-wide distribution, which enable both wide and rapid dissemination of research output.

The books of this series are submitted to indexing to Web of Science, EI-Compendex, DBLP, SCOPUS, Google Scholar and Springerlink.

More information about this series at http://www.springer.com/series/7092

József Dombi · Tamás Jónás

Advances in the Theory of Probabilistic and Fuzzy Data Scientific Methods with Applications

József Dombi
Institute of Informatics
University of Szeged
Szeged, Hungary

Tamás Jónás
Institute of Business Economics
Eötvös Loránd University
Budapest, Hungary

ISSN 1860-949X ISSN 1860-9503 (electronic)
Studies in Computational Intelligence
ISBN 978-3-030-51951-3 ISBN 978-3-030-51949-0 (eBook)
https://doi.org/10.1007/978-3-030-51949-0

This Springer imprint is published by the registered company Springer Nature Switzerland AG
The registered company address is: Gewerbestrasse 11, 6330 Cham, Switzerland

Preface

This monograph focuses on advances in the areas of soft computational and probabilistic methods that we published over the past few years. The text consists of five separate chapters, in which theoretical results and application possibilities are discussed as well. The proposed methods and techniques are connected with various areas of science, engineering and the business economics. It is pointed out how the plausibility and belief measures and the λ-additive and ν-additive measures can be used to model uncertainty. A chapter is dedicated to a probability distribution family which may be viewed as an alternative to four notable probability distributions. This distribution family can be applied in reliability engineering and in economics as well. A fuzzy arithmetic-based time series model is presented with the aim of providing the users with an easy-to-use modeling and forecasting technique. Flexible fuzzy numbers are then proposed for Likert scale-based evaluations. These fuzzy numbers can be successfully applied in performance evaluations where the evaluators need to incorporate vagueness, which may originate from their uncertainty or from the variability of the perceived performance values.

We are confident that the new methods and techniques can be successfully applied to solving a wide range of problems in many areas including engineering, economics, biology, chemistry and the medical sciences.

Szeged, Hungary — József Dombi
Budapest, Hungary — Tamás Jónás

Introduction

Nowadays, application of computer science is common in many areas like engineering, economics, biology, chemistry and the medical sciences. Most of the computational techniques are hybrids as they utilize various branches of mathematics. This monograph summarizes our novel results that were published over the past few years. There are five separate chapters which describe soft computational, probabilistic and hybrid methods. The latter methods may be viewed as synergies between the soft computational and the probabilistic approaches. The monograph—in line with our motivations—besides the theoretical results, also provides the reader with potential applications in the areas of engineering and business economics.

The Chap. 1 may be regarded as a brief review of some important monotone measures; namely, the belief-, probability- and plausibility measures. The practical importance of these measures lies in the fact that human decisions under uncertainty are commonly based upon beliefs and plausibilities associated with possible events. Here, the most important properties of these measures are summarized and the uncertainty modeling capabilities of the belief- and plausibility measures are demonstrated. In this chapter, it is shown how the well-known Ellsberg paradox can be resolved by using belief- and plausibility measures, which may be viewed as lower and upper probability measures, respectively.

Chapter 2 discusses one of the most widely applied classes of monotone measures, the class of λ-additive measures (Sugeno λ-measures). Although many theoretical and practical articles have been published concerning λ-additive measure, its properties and its applicability, there have been no studies that dealt with the general form of λ-additive measure of the union of n sets. With the aim of filling this gap, here, the general Poincaré formula for λ-additive measures is presented. This formula, which may be viewed as a generalization of the well-known Poincaré formula in probability theory, tells us how λ-additive measure of the union of n sets can be computed.

In this chapter, it is also shown how λ-additive measure is connected with the belief-, probability- and plausibility measures. Furthermore, new characteristic inequalities for belief- and plausibility measures using λ-additive measure are presented. It is a well-known fact that λ-additive measure corresponds to a belief-, probability- and plausibility measure, if $\lambda > 0$, $\lambda = 0$ and $\lambda \in (-1, 0)$, respectively. However, since the zero value of λ does not divide the domain $(-1, \infty)$ into two symmetric sub-domains, it is difficult to judge the 'plausibilitiness' or 'beliefness' of a λ-additive measure based on the value of parameter λ. Here, the so-called ν-additive measure, which may be treated as an alternatively parameterized λ-additive measure, is introduced and it is demonstrated that ν parameter has an important semantic meaning. Namely, ν is the fix point of ν-additive complement operation. Also, it is shown that ν-additive measure corresponds to a belief-, probability- and plausibility measure, if $\nu \in (0, 1/2)$, $\nu = 1/2$ and $\nu \in (1/2, 1)$, respectively. Therefore, ν-additive measure can be utilized to characterize the 'plausibilitiness' or 'beliefness' on a normalized scale. Moreover, in this chapter, it is described how λ-additive and ν-additive measures are connected with rough sets, multi-attribute utility functions and with certain operators of fuzzy logic.

A new four-parameter probability distribution function family is introduced and some of its applications are discussed in Chap. 3. This new distribution function is called the pliant probability distribution function, which may be viewed as an alternative to some notable distribution functions. The cumulative distribution function of the novel probability distribution is based on the so-called omega function. It is demonstrated here that both the omega and the exponential function $f(x) = e^{\alpha x^\beta}$ $(x, \alpha, \beta \in \mathbb{R}, \beta > 0)$ may be deduced from a common differential equation that is called the generalized exponential differential equation. Furthermore, it is shown that the omega function, which has parameters α, β and d $(\alpha, \beta, d \in \mathbb{R}, \beta, d > 0)$, is asymptotically identical with the exponential function $f(x) = e^{\alpha x^\beta}$.

Also, it is shown in detail how the pliant probability distribution function can be utilized to approximate some well-known probability distribution functions, the formulas of which include exponential terms. Namely, it is demonstrated that the asymptotic pliant distribution function may coincide with the Weibull, exponential and logistic probability distribution functions. Then, it is shown that with appropriate parameter settings, the pliant probability distribution function can approximate the standard normal probability distribution function quite well, while the approximating formula is very simple and contains only one parameter. Here, it is also noted that this approximation formula may be viewed as a special case of a modifier operator in the continuous-valued logic. The flexibility of this new probability distribution function lays the foundations for its applications in many fields of science and in wide range of modeling problems. It is pointed out that the pliant probability distribution function, as an alternative to the Weibull- and exponential distribution functions, can be utilized for modeling constant, monotonic and bathtub-shaped hazard functions in reliability theory. Furthermore, it is demonstrated that a function transformed from the new probability distribution function can be applied in the so-called kappa regression analysis method, which may be viewed as an alternative to logistic regression.

In Chap. 4, a new fuzzy time series modeling technique is proposed. This technique is based on fuzzy inference methods in which the fuzzy output is obtained by fuzzy arithmetic operations; namely, via the weighted summation of pliant numbers that are the consequents of fuzzy rules. This fuzzy inference method is called the Pliant Arithmetic-Based Fuzzy Inference System (PAFIS). The defuzzification method of the system has two notable advantages. Firstly, it does not require any numerical integration to generate the crisp output; and secondly, it runs in a constant time. Afterwards, it is explained how the Pliant Arithmetic-Based Fuzzy Time Series (PAFTS) model can be derived by utilizing the above-mentioned inference methods.

The greatest advantages of our PAFTS model lies in its simplicity and easy-to-use characteristics. Once the fuzzy rule consequents are obtained, the time series modeling and forecasting become very simple. In order to evaluate the forecasting performance of the PAFTS method, 17 time series were analyzed including the Australian Beer Consumption (ABC) from 1956 to 1994, the Istanbul Stock Exchange Market (BIST100) Index for the time period between 2009 and 2013, and the Taiwan Stock Exchange Capitalization Weighted Stock Index (TAIEX) for the time period between 1999 and 2004. The forecasting results of the proposed methods were compared with the results of various well-known methods. Based on our empirical results, the PAFTS method may be viewed as novel viable time series modeling technique.

In Chap. 5, the so-called flexible fuzzy number, the membership function of which may have various shapes depending on its shape parameter λ is introduced. Namely, this membership function can be bell-shaped $(\lambda > 1)$, triangular $(\lambda = 1)$, or 'reverse' bell-shaped $(0 < \lambda < 1)$. This feature of the flexible fuzzy number allows us to express the soft equation 'x is approximately equal to x_0' in various ways.

Here, it is demonstrated that if the shape parameter λ is fixed, then the set of flexible fuzzy numbers is closed under the multiplication by scalar, fuzzy addition and weighted average operations. The pliancy of flexible fuzzy numbers and the above-mentioned properties of the operations over them make these numbers suitable for Likert scale-based fuzzy evaluations. On the one hand, they can be used to deal with the vagueness that may originate from the uncertainty of the evaluators or from the variability of the perceived performance values. On the other hand, in a multi-dimensional Likert scale-based evaluation, having the rating result in each evaluation dimension represented by a flexible fuzzy number, these results can be readily aggregated into one flexible fuzzy number using the above-mentioned arithmetic operations.

Next, as a generalization of the flexible fuzzy numbers the extended flexible fuzzy numbers, which may have different left hand side and right hand side shape parameters, are introduced. Also, it is demonstrated that the asymptotic flexible fuzzy number is just a quasi fuzzy number composed of an increasing left hand side and a decreasing right hand side sigmoid function. Exploiting this property of the extended flexible fuzzy numbers allows us to perform approximate fuzzy arithmetic operations over them in a simple way. Owing to their advantageous properties both of the flexible fuzzy numbers and the extended flexible fuzzy numbers, the proposed fuzzy evaluation methods can be readily implemented and utilized in practice.

Contents

List of Figures

List of Tables

Chapter 1
Belief, Probability and Plausibility

There are many situations in our lives where we need to make decisions and act under uncertainty. Although probability theory may be viewed as a standard approach to modeling uncertainty, there are situations where it is not the best approach to representing information. With the aim of filling this gap, over the past decades, various generalizations and alternatives have been developed for probability theory. Some notable approaches and theories are the Dempster–Shafer theory of evidence [1, 2], the plausibility measures [3], the possibility measures [4], and the subjective probability [5]. In practice, human decisions under uncertainty are commonly based upon beliefs and plausibilities associated with possible events. Therefore, there has been a great interest in belief- and plausibility measures and in their applications (see, e.g. [6–10]). In this chapter, following Dempster–Shafer's theory of evidence, we will give a brief overview of the belief-, probability- and plausibility measures. This chapter is connected with the following publications of ours: [11, 12].

From now on, we will use the common notations $\cap$, $\cup$ and $\setminus$ for the intersection, union and difference operations over sets, respectively. Also, will use the notation $\overline{A}$ for the complement of set A.

1.1 Monotone Measures

Now, we will introduce the monotone measures and give a short overview of them that covers the probability-, belief- and plausibility measures.

Definition 1.1 Let Σ be a σ-algebra on the set X. Then the function $\mu\colon \Sigma \to [0, 1]$ is a monotone measure on the measurable space (X, Σ) if and only if μ satisfies the following requirements:

J. Dombi and T. Jónás, *Advances in the Theory of Probabilistic and Fuzzy Data Scientific Methods with Applications*, Studies in Computational Intelligence 814,
https://doi.org/10.1007/978-3-030-51949-0_1

(1) $\mu(\emptyset) = 0$, $\mu(X) = 1$
(2) if $B \subseteq A$, then $\mu(B) \leq \mu(A)$ for any $A, B \in \Sigma$ (monotonicity)
(3) if $\forall i \in \mathbb{N}$, $A_i \in \Sigma$ and (A_i) is monotonic ($A_1 \subseteq A_2 \subseteq \cdots \subseteq A_n \subseteq \cdots$ or $A_1 \supseteq A_2 \supseteq \cdots \supseteq A_n \cdots$), then $\lim_{i\to\infty} \mu(A_i) = \mu\left(\lim_{i\to\infty} A_i\right)$ (continuity).

If X is a finite set, then the continuity requirement in Definition 1.1 can be disregarded and the monotone measure is defined as follows.

Definition 1.2 The function $\mu\colon \mathcal{P}(X) \to [0, 1]$ is a monotone measure on the finite set X if and only if μ satisfies the following requirements:

(1) $\mu(\emptyset) = 0$, $\mu(X) = 1$
(2) if $B \subseteq A$, then $\mu(B) \leq \mu(A)$ for any $A, B \in \mathcal{P}(X)$ (monotonicity).

Note that the monotone measures given by Definitions 1.1 and 1.2 are known as fuzzy measures, which were originally defined by Choquet [13] and Sugeno [14].

1.2 Probability Measure

Definition 1.3 Let Σ be a σ-algebra over the set X. Then the function $Pr\colon \Sigma \to [0, 1]$ is a probability measure on the space (X, Σ) if and only if Pr satisfies the following requirements:

(1) $\forall A \in \Sigma\colon Pr(A) \geq 0$
(2) $Pr(X) = 1$
(3) $\forall A_1, A_2, \ldots, \in \Sigma$, if $A_i \cap A_j = \emptyset, \forall i \neq j$, then

$$Pr\left(\bigcup_{i=1}^{\infty} A_i\right) = \sum_{i=1}^{\infty} Pr(A_i).$$

Remark 1.1 If X is a finite set, then requirement (3) in Definition 1.3 can be reduced to the following requirement: for any disjoint $A, B \in \mathcal{P}(X)$, $Pr(A \cup B) = Pr(A) + Pr(B)$.

Later on, we will use the well-known Poincaré formula of probability theory:

$$Pr\left(\bigcup_{i=1}^{n} A_i\right) = \sum_{k=1}^{n} (-1)^{k-1} \sum_{1\leq i_1 < \cdots < i_k \leq n} Pr\left(A_{i_1} \cap \cdots \cap A_{i_k}\right), \tag{1.1}$$

where Pr is a probability measure on X and $A_1, \ldots, A_n \in \mathcal{P}(X)$.

1.3 Belief Measure and Plausibility Measure

Definition 1.4 The function $Bl\colon \mathcal{P}(X) \to [0, 1]$ is a belief measure on the finite set X, if and only if Bl satisfies the following requirements:

(1) $Bl(\emptyset) = 0$, $Bl(X) = 1$
(2) for any $A_1, A_2, \ldots, A_n \in \mathcal{P}(X)$,

$$Bl(A_1 \cup A_2 \cup \cdots \cup A_n) \geq$$
$$\geq \sum_{k=1}^{n} \sum_{1 \leq i_1 < i_2 \cdots < i_k \leq n} (-1)^{k-1} Bl\left(A_{i_1} \cap A_{i_2} \cap \cdots \cap A_{i_k}\right). \tag{1.2}$$

Here, $Bl(A)$ is interpreted as the grade of belief that a given element of X belongs to A.

Lemma 1.1 *If Bl is a belief measure on the finite set X, then for any $A \in \mathcal{P}(X)$,*

$$Bl(A) + Bl(\overline{A}) \leq 1.$$

Proof Noting Definition 1.4, we have

$$1 = Bl(A \cup \overline{A}) \geq Bl(A) + Bl(\overline{A}) - Bl(A \cap \overline{A}) = Bl(A) + Bl(\overline{A}). \qquad \square$$

The inequality $Bl(A) + Bl(\overline{A}) \leq 1$ means that a lack of belief in A does not imply a strong belief in $\overline{A}$. In particular, total ignorance is modeled by the belief function Bl_i such that $Bl_i(A) = 0$ if $A \neq X$ and $Bl_i(A) = 1$ if $A = X$. The following proposition is about the monotonicity of belief measures.

Proposition 1.1 *If X is a finite set, Bl is a belief measure on X, $A, B \in \mathcal{P}(X)$ and $B \subseteq A$, then $Bl(B) \leq Bl(A)$.*

Proof Let $B \subseteq A$. Hence, there exists a $C \in \mathcal{P}(X)$ such that $A = B \cup C$ and $B \cap C = \emptyset$. Now, by utilizing the definition of the belief measure and the fact that $B \cap C = \emptyset$, we get

$$Bl(A) = Bl(B \cup C) \geq Bl(B) + Bl(C) \geq Bl(B). \qquad \square$$

Corollary 1.1 *The belief measure given by Definition 1.4 is a monotone measure.*

Proof Let Bl be a belief measure. It follows from Definition 1.4 that Bl satisfies criterion (1) for a monotone measure given in Definition 1.2. Moreover, the monotonicity of Bl was proven in Proposition 1.1; that is, Bl also satisfies criterion (2) in Definition 1.2. $\square$

Definition 1.5 The function $Pl : \mathcal{P}(X) \to [0, 1]$ is a plausibility measure on the finite set X, if and only if Pl satisfies the following requirements:

(1) $Pl(\emptyset) = 0$, $Pl(X) = 1$
(2) for any $A_1, A_2, \ldots, A_n \in \mathcal{P}(X)$,

$$Pl(A_1 \cap A_2 \cap \cdots \cap A_n) \leq$$
$$\leq \sum_{k=1}^{n} \sum_{1 \leq i_1 < i_2 \cdots < i_k \leq n} (-1)^{k-1} Pl\left(A_{i_1} \cup A_{i_2} \cdots \cup A_{i_k}\right). \tag{1.3}$$

Lemma 1.2 *If Pl is a plausibility measure on the finite set X, then for any $A \in \mathcal{P}(X)$,*

$$Pl(A) + Pl(\overline{A}) \geq 1.$$

Proof Noting Definition 1.5, we have

$$0 = Pl(A \cap \overline{A}) \leq Pl(A) + Pl(\overline{A}) - Pl(A \cup \overline{A}) = Pl(A) + Pl(\overline{A}) - 1,$$

from which $Pl(A) + Pl(\overline{A}) \geq 1$ follows. □

Here, $Pl(A)$ is interpreted as the plausibility of the event A. The following proposition is about the monotonicity of plausibility measures.

Proposition 1.2 *If X is a finite set, Pl is a plausibility measure on X, $A, B \in \mathcal{P}(X)$ and $B \subseteq A$, then $Pl(B) \leq Pl(A)$.*

Proof Let $B \subseteq A$. Let $C \in \mathcal{P}(X)$ such that $A \cap C = B$ and $A \cup C = X$. Now, by utilizing the definition of plausibility measure, and the fact that $A \cup C = X$ and $Pl(C) \leq 1$, we get

$$Pl(B) = Pl(A \cap C) \leq Pl(A) + Pl(C) - Pl(A \cup C) =$$
$$= Pl(A) + Pl(C) - 1 \leq Pl(A).$$
□

Corollary 1.2 *The plausibility measure given by Definition 1.5 is a monotone measure.*

Proof Let Pl be a plausibility measure. It follows from Definition 1.5 that Pl satisfies criterion (1) for a monotone measure given in Definition 1.2. Next, the monotonicity of Pl was proven in Proposition 1.2; that is, Pl also satisfies criterion (2) in Definition 1.2.

Remark 1.2 It immediately follows from the definitions of belief- and plausibility measures and from the Poincaré formula of probability theory in Eq. (1.1) that if μ is a probability measure on the finite set X, then μ is also a belief- and plausibility measure on X.

We will later utilize the following lemma.

Lemma 1.3

$$\sum_{k=0}^{n} \binom{n}{k} (-1)^k = 0. \tag{1.4}$$

Proof By using the binomial theorem, we immediately get that

$$0 = (1-1)^n = \sum_{k=0}^{n} \binom{n}{k} (-1)^k.$$

□

The following proposition states an interesting connection between the belief measure and the plausibility measure.

Proposition 1.3 *Let X be a finite set and let $\mu_1, \mu_2 \colon \mathcal{P}(X) \to [0, 1]$ be two monotone measures on X such that*

$$\mu_2(A) = 1 - \mu_1(\overline{A}) \tag{1.5}$$

holds for any $A \in \mathcal{P}(X)$. Then, either (1) μ_1 is a belief measure on X if and only if μ_2 is a plausibility measure on X, or (2) μ_1 is a plausibility measure on X if and only if μ_2 is a belief measure on X.

Proof We will prove case (1), and the proof of case (2) is similar. Firstly, we will show that if μ_1 is a belief measure on X and $\mu_2(A)$ is given as $\mu_2(A) = 1 - \mu_1(\overline{A})$ for any $A \in \mathcal{P}(X)$, then μ_2 is a plausibility measure on X. Let μ_1 be a belief measure on X and $\mu_2(A) = 1 - \mu_1(\overline{A})$ for any $A \in \mathcal{P}(X)$. Then, $\mu_2(\emptyset) = 0$ and $\mu_2(X) = 1$ trivially follow from the fact that μ_1 is a belief measure and $\mu_2(A) = 1 - \mu_1(\overline{A})$. That is, function μ_2 satisfies requirement (1) for a plausibility measure given in Definition 1.5. Furthermore, since function μ_1 is a belief measure, the inequality

$$\mu_1(A_1 \cup A_2 \cup \cdots \cup A_n) \geq$$
$$\geq \sum_{k=1}^{n} \sum_{1 \leq i_1 < i_2 \cdots < i_k \leq n} (-1)^{k-1} \mu_1 \left(A_{i_1} \cap A_{i_1} \cap \cdots \cap A_{i_k}\right). \tag{1.6}$$

holds for any $A_1, A_2, \ldots, A_n \in \mathcal{P}(X)$. From the condition $\mu_2(A) = 1 - \mu_1(\overline{A})$, we also have that $\mu_1(A) = 1 - \mu_2(\overline{A})$. Next, applying the inequality in Eq. (1.6) to the complement sets $\overline{A}_1, \overline{A}_2, \ldots, \overline{A}_n \in \mathcal{P}(X)$ and utilizing the fact that $\mu_1(A) = 1 - \mu_2(\overline{A})$, we get

$$
\begin{aligned}
1-\mu_2(\overline{\overline{A}_1\cup\overline{A}_2\cup\cdots\cup\overline{A}_n}) \geq\ & 1-\mu_2(\overline{\overline{A}}_1)+1-\mu_2(\overline{\overline{A}}_2)+\cdots+1-\mu_2(\overline{\overline{A}}_n)- \\
& -(1-\mu_2(\overline{\overline{A}_1\cap\overline{A}_2}))-\cdots-(1-\mu_2(\overline{\overline{A}_{n-1}\cap\overline{A}_n}))+\cdots \\
& \cdots+(-1)^{n+1}\left(1-\mu_2(\overline{\overline{A}_1\cap\overline{A}_2\cap\cdots\cap\overline{A}_n})\right)= \\
& =-\mu_2(\overline{\overline{A}}_1)-\mu_2(\overline{\overline{A}}_2)-\cdots-\mu_2(\overline{\overline{A}}_n)+ \\
& +\mu_2(\overline{\overline{A}_1\cap\overline{A}_2})+\cdots+\mu_2(\overline{\overline{A}_{n-1}\cap\overline{A}_n})+\cdots \\
& \cdots+(-1)^n\mu_2(\overline{\overline{A}_1\cap\overline{A}_2\cap\cdots\cap\overline{A}_n})+\binom{n}{1}-\binom{n}{2}+\cdots+\binom{n}{n}(-1)^{n+1}.
\end{aligned}
$$

Noting Eq. (1.4), the previous inequality can be written as

$$
\begin{aligned}
1-\mu_2(\overline{\overline{A}_1\cup\overline{A}_2\cup\cdots\cup\overline{A}_n}) \geq 1-\mu_2(\overline{\overline{A}}_1)-\mu_2(\overline{\overline{A}}_2)-\cdots-\mu_2(\overline{\overline{A}}_n)+ \\
+\mu_2(\overline{\overline{A}_1\cap\overline{A}_2})+\cdots+\mu_2(\overline{\overline{A}_{n-1}\cap\overline{A}_n})+\cdots+(-1)^n\mu_2(\overline{\overline{A}_1\cap\overline{A}_2\cap\cdots\cap\overline{A}_n}).
\end{aligned}
$$

Now, applying the De Morgan law to the last inequality, we get

$$
\begin{aligned}
\mu_2(A_1\cap A_2\cap\cdots\cap A_n) \leq \mu_2(A_1)+\mu_2(A_2)+\cdots+\mu_2(A_n)- \\
-\mu_2(A_1\cup A_2)-\cdots-\mu_2(A_{n-1}\cup A_n)+(-1)^n\mu_2(A_1\cup A_2\cup\cdots\cup A_n)= \\
=\sum_{k=1}^{n}\sum_{1\leq i_1<i_2\cdots<i_k\leq n}(-1)^{k-1}\mu_2\left(A_{i_1}\cup A_{i_1}\cup\cdots\cup A_{i_k}\right),
\end{aligned}
$$

which means that function μ_2 is a plausibility measure.

Secondly, we will demonstrate that if μ_2 is a plausibility measure on X and $\mu_2(A)$ is given as $\mu_2(A)=1-\mu_1(\overline{A})$ for any $A\in\mathcal{P}(X)$, then μ_1 is a belief measure on X. Let μ_2 be a plausibility measure on X and $\mu_2(A)=1-\mu_1(\overline{A})$ for any $A\in\mathcal{P}(X)$. These conditions trivially imply that $\mu_1(\emptyset)=0$ and $\mu_1(X)=1$; that is, function μ_1 satisfies requirement (1) for a belief measure given in Definition 1.4. Next, because function μ_2 is a plausibility measure, the inequality

$$
\begin{aligned}
\mu_2(A_1\cap A_2\cap\cdots\cap A_n) \leq \\
\leq\sum_{k=1}^{n}\sum_{1\leq i_1<i_2\cdots<i_k\leq n}(-1)^{k-1}\mu_2\left(A_{i_1}\cup A_{i_2}\cdots\cup A_{i_k}\right)
\end{aligned}
\tag{1.7}
$$

holds for any $A_1, A_2, \ldots, A_n\in\mathcal{P}(X)$. Then, applying the inequality in Eq. (1.7) to the complement sets $\overline{A}_1, \overline{A}_2, \ldots, \overline{A}_n\in\mathcal{P}(X)$ and utilizing the condition that $\mu_2(A)=1-\mu_1(\overline{A})$, we get

$$
\begin{aligned}
1-\mu_1(\overline{\overline{A}_1\cap\overline{A}_2\cap\cdots\cap\overline{A}_n}) &\le 1-\mu_1(\overline{\overline{A}_1})+1-\mu_1(\overline{\overline{A}_2})+\cdots+1-\mu_1(\overline{\overline{A}_n})-\\
&-(1-\mu_1(\overline{\overline{A}_1\cup\overline{A}_2}))-\cdots-(1-\mu_1(\overline{\overline{A}_{n-1}\cup\overline{A}_n}))+\cdots\\
&\cdots+(-1)^{n+1}\left(1-\mu_1(\overline{\overline{A}_1\cup\overline{A}_2\cup\cdots\cup\overline{A}_n})\right)=\\
&=-\mu_1(\overline{\overline{A}_1})-\mu_1(\overline{\overline{A}_2})-\cdots-\mu_1(\overline{\overline{A}_n})+\\
&+\mu_1(\overline{\overline{A}_1\cup\overline{A}_2})+\cdots+\mu_1(\overline{\overline{A}_{n-1}\cup\overline{A}_n})+\cdots\\
&\cdots+(-1)^n\mu_1(\overline{\overline{A}_1\cup\overline{A}_2\cup\cdots\cup\overline{A}_n})+\binom{n}{1}-\binom{n}{2}+\cdots+\binom{n}{n}(-1)^{n+1}.
\end{aligned}
$$

Again, taking into account Eq. (1.4), the previous inequality can be written as

$$
\begin{aligned}
1-\mu_1(\overline{\overline{A}_1\cap\overline{A}_2\cap\cdots\cap\overline{A}_n}) \le 1-\mu_1(\overline{\overline{A}_1})-\mu_1(\overline{\overline{A}_2})-\cdots-\mu_1(\overline{\overline{A}_n})+\\
+\mu_1(\overline{\overline{A}_1\cup\overline{A}_2})+\cdots+\mu_1(\overline{\overline{A}_{n-1}\cup\overline{A}_n})+\cdots+(-1)^n\mu_1(\overline{\overline{A}_1\cup\overline{A}_2\cup\cdots\cup\overline{A}_n}).
\end{aligned}
$$

Now, applying the De Morgan law to the last inequality, we get

$$
\begin{aligned}
\mu_1(A_1\cup A_2\cup\cdots\cup A_n) \ge \mu_1(A_1)+\mu_1(A_2)+\cdots+\mu_1(A_n)-\\
-\mu_1(A_1\cap A_2)-\cdots-\mu_1(A_{n-1}\cap A_n)+(-1)^n\mu_1(A_1\cap A_2\cap\cdots\cap A_n)=\\
=\sum_{k=1}^{n}\sum_{1\le i_1<i_2\cdots<i_k\le n}(-1)^{k-1}\mu_1\left(A_{i_1}\cap A_{i_1}\cap\cdots\cap A_{i_k}\right).
\end{aligned}
$$

Hence, μ_1 is a belief measure. □

Later, we will use the concept of a dual pair of belief- and plausibility measures.

Definition 1.6 Let Bl and Pl be a belief measure and a plausibility measure, respectively, on set X. Then Bl and Pl are said to be a dual pair of belief- and plausibility measures if and only if

$$Pl(A)=1-Bl(\overline{A})$$

holds for any $A\in\mathcal{P}(X)$.

Remark 1.3 The monotonicity of the plausibility measure Pl can also be demonstrated by utilizing the duality $Pl(A)=1-Bl(\overline{A})$ and the monotonicity of the belief measure Bl. Namely, if $B\subseteq A$, then $\overline{A}\subseteq\overline{B}$ and so

$$Bl(\overline{A})\le Bl(\overline{B}),$$

from which

$$1-Bl(\overline{A})\ge 1-Bl(\overline{B}),$$

which means that

$$Pl(A)\ge Pl(B).$$

1.3.1 *Belief and Plausibility in Dempster–Shafer Theory*

In the Dempster–Shafer theory of evidence, a belief mass is assigned to each element of the power set $\mathcal{P}(X)$, where X is a finite set. The belief mass is given by the so-called basic probability assignment m from $\mathcal{P}(X)$ to $[0, 1]$ that is defined as follows.

Definition 1.7 The function $m: \mathcal{P}(X) \to [0, 1]$ is a basic probability assignment (mass function) on the finite set X, if and only if m satisfies the following requirements:

(1) $m(\emptyset) = 0$
(2) $\sum_{A \in \mathcal{P}(X)} m(A) = 1$.

The subsets A of X for which $m(A) > 0$ are called the focal elements of m. Let $x \in A$ and $A \in \mathcal{P}(X)$. Then, the mass $m(A)$ can be interpreted as the probability of knowing $x \in A$ given the available evidence. Utilizing a given basic probability assignment m, in the Dempster–Shafer theory, the belief- and plausibility measures are defined as follows.

Definition 1.8 The function $Bl: \mathcal{P}(X) \to [0, 1]$ is a belief measure on the finite set X, if and only if for any $A \in \mathcal{P}(X)$

$$Bl(A) = \sum_{B|B \subseteq A} m(B),$$

where $m: \mathcal{P}(X) \to [0, 1]$ is a basic probability assignment (mass function) on the set X.

Definition 1.9 The function $Pl: \mathcal{P}(X) \to [0, 1]$ is a plausibility measure on the finite set X, if and only if for any $A \in \mathcal{P}(X)$

$$Pl(A) = \sum_{B|B \cap A \neq \emptyset} m(B),$$

where $m: \mathcal{P}(X) \to [0, 1]$ is a basic probability assignment (mass function) on the set X.

The belief measure $Bl(A)$ and the plausibility measure $Pl(A)$ of the set A can be interpreted as follows:

- $Bl(A)$ is the degree to which the available information (evidence) supports A.
- $Pl(A)$ is an upper bound on the degree to which the available information (evidence) could support A.

The following lemma is about the duality of the belief- and plausibility measures given in Definitions 1.8 and 1.9.

Lemma 1.4 *Let X be a finite set. If the functions Bl, $Pl\colon \mathcal{P}(X) \to [0, 1]$ are given as*

$$Bl(A) = \sum_{B|B \subseteq A} m(B),$$

$$Pl(A) = \sum_{B|B \cap A \neq \emptyset} m(B),$$

for any $A \in \mathcal{P}(X)$, where $m\colon \mathcal{P}(X) \to [0, 1]$ is a basic probability assignment (mass function) on the set X, then

$$Pl(A) + Bl(\overline{A}) = 1$$

holds for any $A \in \mathcal{P}(X)$.

Proof Let $A \in \mathcal{P}(X)$ and let $m\colon \mathcal{P}(X) \to [0, 1]$ be a basic probability assignment on the universe X. Furthermore, let the sets S_{Bl} and S_{Pl} be given by

$$S_{Bl} = \{B|B \subseteq \overline{A}\}$$

$$S_{Pl} = \{B|B \cap A \neq \emptyset\}.$$

On the one hand, since m is a basic probability assignment on X,

$$\sum_{S \in \mathcal{P}(X)} m(S) = 1.$$

On the other hand, it can be seen that $S_{Bl} \cup S_{Pl} = X$ and $S_{Bl} \cap S_{Pl} = \emptyset$. Thus,

$$1 = \sum_{S \in \mathcal{P}(X)} m(S) = \sum_{S \in S_{Pl}} m(S) + \sum_{S \in S_{Bl}} m(S) = Pl(A) + Bl(\overline{A}).$$ □

Note that, in line with this result, the plausibility of a subset A of the finite set X was defined by Shafer [2] as

$$Pl(A) = 1 - Bl(\overline{A}),$$

where Bl is a belief function.

Corollary 1.3 *If Bl, $Pl\colon \mathcal{P}(X) \to [0, 1]$ are two monotone measures on the finite set X, $Pl(A)$ is given by*

$$Pl(A) = \sum_{B|B \cap A \neq \emptyset} m(B),$$

for any $A \in \mathcal{P}(X)$, where $m\colon \mathcal{P}(X) \to [0, 1]$ is a basic probability assignment on the set X, and

$$Pl(A) + Bl(\overline{A}) = 1$$

holds for any $A \in \mathcal{P}(X)$*, then the equation*

$$Bl(A) = \sum_{B|B \subseteq A} m(B)$$

holds for any $A \in \mathcal{P}(X)$ *as well.*

Proof Here, we will give an indirect proof for the proposition of this corollary. Let us assume that all the conditions of the corollary are satisfied, and there exists an $A \in \mathcal{P}(X)$ such that

$$Bl(A) \neq \sum_{B|B \subseteq A} m(B).$$

Now, we will show that this assumption leads to a contradiction. Let $Bl^*(A)$ be given by

$$Bl^*(A) = \sum_{B|B \subseteq A} m(B).$$

Here, on the one hand, based on Lemma 1.4, the equation $Pl(A) + Bl^*(\overline{A}) = 1$ holds. On the other hand, since $Bl(A) \neq Bl^*(A)$, $Bl(\overline{A}) \neq Bl^*(\overline{A})$ holds, we have $Pl(A) + Bl(\overline{A}) \neq 1$, which contradicts the requirement that $Pl(A) + Bl(\overline{A}) = 1$.□

Corollary 1.4 *If* $Bl, Pl \colon \mathcal{P}(X) \to [0, 1]$ *are two monotone measures on the finite set* X*,* $Bl(A)$ *is given by*

$$Bl(A) = \sum_{B|B \subseteq A} m(B),$$

for any $A \in \mathcal{P}(X)$*, where* $m \colon \mathcal{P}(X) \to [0, 1]$ *is a basic probability assignment on the set* X*, and*

$$Pl(A) + Bl(\overline{A}) = 1$$

holds for any $A \in \mathcal{P}(X)$*, then the equation*

$$Pl(A) = \sum_{B|B \cap A \neq \emptyset} m(B)$$

holds for any $A \in \mathcal{P}(X)$ *as well.*

Proof This corollary can be proved by contradiction in a way similar to the proof of Corollary 1.3. □

We will use of the following lemma.

Lemma 1.5 (Lemma 4.1 in *[15]*) *If A is a nonempty finite set, then*

$$\sum_{B \subseteq A} (-1)^{|B|} = 0. \tag{1.8}$$

Proof Let $A = \{x_1, x_2, \ldots, x_n\}$. Then, by noting Lemma 1.3, we have

$$\sum_{B \subseteq A} (-1)^{|B|} = \sum_{k=0}^{n} \binom{n}{k} (-1)^k = 0.$$

□

Now, we will demonstrate a key property of the belief measure given in Definition 1.8 that we will utilize later.

Proposition 1.4 *If $m\colon \mathcal{P}(X) \to [0, 1]$ is a basic probability assignment on the finite set X and for any $A \in \mathcal{P}(X)$*

$$Bl(A) = \sum_{B | B \subseteq A} m(B), \tag{1.9}$$

(that is, Bl satisfies the requirement for a belief measure given in Definition 1.8), then the inequality

$$Bl(A_1 \cup A_2 \cup \cdots \cup A_n) \geq \geq \sum_{k=1}^{n} \sum_{1 \leq i_1 < i_2 < \cdots < i_k \leq n} (-1)^{k-1} Bl\left(A_{i_1} \cap A_{i_1} \cap \cdots \cap A_{i_k}\right) \tag{1.10}$$

holds for any $A_1, A_2, \ldots, A_n \in \mathcal{P}(X)$.

Proof (*See also the proof of Theorem 4.13 in* [15]) Notice that the right hand side of Eq. (1.10) can be written as

$$\sum_{k=1}^{n} \sum_{1 \leq i_1 < i_2 < \cdots < i_k \leq n} (-1)^{k-1} Bl\left(A_{i_1} \cap A_{i_1} \cap \cdots \cap A_{i_k}\right) = = \sum_{\substack{I \subseteq \{1,\ldots,n\} \\ I \neq \emptyset}} (-1)^{|I|+1} Bl\left(\bigcap_{i \in I} A_i\right) = = \sum_{\substack{I \subseteq \{1,\ldots,n\} \\ I \neq \emptyset}} \left((-1)^{|I|+1} \sum_{B \subseteq \bigcap_{i \in I} A_i} m(B) \right). \tag{1.11}$$

For any $B \in \mathcal{P}(X)$, let the index set $I(B)$ be defined by

$$I(B) = \{i | B \subseteq A_i, 1 \leq i \leq n\}.$$

Then, the chain in Eq. (1.11) can be continued as

$$\sum_{\substack{I\subseteq\{1,\ldots,n\}\\ I\neq\emptyset}}\left((-1)^{|I|+1}\sum_{B\subseteq\cap_{i\in I}A_i} m(B)\right)=$$

$$=\sum_{B|I(B)\neq\emptyset}\left(m(B)\sum_{\substack{I\subseteq I(B)\\ I\neq\emptyset}}(-1)^{|I|+1}\right)= \tag{1.12}$$

$$=\sum_{B|I(B)\neq\emptyset}\left(m(B)\left(1-\sum_{I\subseteq I(B)}(-1)^{|I|}\right)\right).$$

Now, by taking into account Lemma 1.5, we have

$$\sum_{I\subseteq I(B)}(-1)^{|I|}=0.$$

Hence, the chain in Eq. (1.12) can be continued as

$$\sum_{B|I(B)\neq\emptyset}\left(m(B)\left(1-\sum_{I\subseteq I(B)}(-1)^{|I|}\right)\right)=$$
$$=\sum_{B|I(B)\neq\emptyset} m(B)\leq\sum_{B\subseteq\cup_{i=1}^{n}A_i} m(B). \tag{1.13}$$

Next, by noting Eqs. (1.9), (1.11), (1.12) and (1.13), we have

$$\sum_{k=1}^{n}\sum_{1\leq i_1<i_2<\cdots<i_k\leq n}(-1)^{k-1}Bl\left(A_{i_1}\cap A_{i_1}\cap\cdots\cap A_{i_k}\right)\leq$$
$$\leq\sum_{B\subseteq\cup_{i=1}^{n}A_i} m(B)=Bl(A_1\cup A_2\cup\cdots\cup A_n).$$

□

We will utilize the following proposition later on.

Proposition 1.5 *Let μ_1 and μ_2 be two set functions on the finite set X. Then, for any $A\in\mathcal{P}(X)$*

$$\mu_1(A)=\sum_{B|B\subseteq A}\mu_2(B) \tag{1.14}$$

holds if and only if

$$\mu_2(A)=\sum_{B|B\subseteq A}(-1)^{|A|-|B|}\mu_1(B) \tag{1.15}$$

holds.

Proof Firstly, we will show that if Eq. (1.14) holds, then Eq. (1.15) holds as well. Let $A \in \mathcal{P}(X)$. Here, based on the condition that Eq. (1.14) holds for any $A \in \mathcal{P}(X)$, we have the equation

$$\mu_1(A') = \sum_{B|B\subseteq A'} \mu_2(B)$$

for every $A' \subseteq A$. Now, multiplying both sides of each of these equations by $(-1)^{|A|-|A'|}$ and summing them results

$$\sum_{A'|A'\subseteq A} (-1)^{|A|-|A'|}\mu_1(A') = \sum_{A'\subseteq A}\sum_{B\subseteq A'}(-1)^{|A'|-|B|}\mu_2(B) =$$

$$= \left(\binom{|A|-1}{0}(-1)^0 + \binom{|A|-1}{1}(-1)^1 + \cdots \right.$$

$$\left. \cdots + \binom{|A|-1}{|A|-1}(-1)^{|A|-1}\right)\sum_{|B|=1}\mu_2(B) +$$

$$+\left(\binom{|A|-2}{0}(-1)^0 + \binom{|A|-2}{1}(-1)^1 + \cdots \right.$$

$$\left. \cdots + \binom{|A|-2}{|A|-2}(-1)^{|A|-2}\right)\sum_{|B|=2}\mu_2(B) + \cdots$$

$$\cdots + \left(\binom{1}{0}(-1)^0\binom{1}{1}(-1)^1\right)\sum_{|B|=|A|-1}\mu_2(B) + \mu_2(A).$$

And, noting Eq. (1.4), from the previous equation we have

$$\mu_2(A) = \sum_{A'|A'\subseteq A} (-1)^{|A|-|A'|}\mu_1(A'),$$

which is identical to Eq. (1.15).

Secondly, we will show that if Eq. (1.15) holds for $A \in \mathcal{P}(X)$, then Eq. (1.14) holds for A as well. Let $A \in \mathcal{P}(X)$. Here, utilizing the condition that Eq. (1.15) holds for any $A \in \mathcal{P}(X)$, we have the equation

$$\mu_2(A') = \sum_{B|B\subseteq A'} (-1)^{|A'|-|B|}\mu_1(B)$$

for every $A' \subseteq A$. Now, by summing these equations we get

$$\sum_{A'|A'\subseteq A} \mu_2(A') = \sum_{A'\subseteq A} \sum_{B\subseteq A'} (-1)^{|A'|-|B|} \mu_1(B) =$$

$$= \left(\binom{|A|-1}{0}(-1)^0 + \binom{|A|-1}{1}(-1)^1 + \cdots \right.$$

$$\left. \cdots + \binom{|A|-1}{|A|-1}(-1)^{|A|-1} \right) \sum_{|B|=1} \mu_1(B) +$$

$$+ \left(\binom{|A|-2}{0}(-1)^0 + \binom{|A|-2}{1}(-1)^1 + \cdots \right.$$

$$\left. \cdots + \binom{|A|-2}{|A|-2}(-1)^{|A|-2} \right) \sum_{|B|=2} \mu_1(B) + \cdots$$

$$\cdots + \left(\binom{1}{0}(-1)^0 \binom{1}{1}(-1)^1 \right) \sum_{|B|=|A|-1} \mu_1(B) + \mu_1(A).$$

Next, making the use of Eq. (1.4), from the previous equation we get

$$\mu_1(A) = \sum_{A'|A'\subseteq A} \mu_2(A'),$$

which is identical to Eq. (1.14). □

Corollary 1.5 *If $Bl\colon \mathcal{P}(X) \to [0, 1]$ is a set function on the finite set X, Bl satisfies the requirements for a belief measure given in Definition 1.8 and $m\colon \mathcal{P}(X) \to [0, 1]$ is the basic probability assignment function of Bl, then*

$$m(A) = \sum_{B\subseteq A} (-1)^{|A|-|B|} Bl(B)$$

holds for any $A \in \mathcal{P}(X)$.

Proof The corollary immediately follows from Proposition 1.5. □

This corollary tells us that a basic probability assignment can be represented by its belief function.

The following theorem demonstrates that the two definitions for belief measure given in Definition 1.4 and in Definition 1.8 are equivalent.

Theorem 1.1 *Let X be a finite set and let $Bl\colon \mathcal{P}(X) \to [0, 1]$ be a measure on X such that $Bl(\emptyset) = 0$ and $Bl(X) = 1$. Then, for any $A_1, A_2, \ldots, A_n \in \mathcal{P}(X)$, the inequality*

$$Bl(A_1 \cup A_2 \cup \cdots \cup A_n) \geq$$

$$\geq \sum_{k=1}^{n} \sum_{1\leq i_1<i_2<\cdots<i_k\leq n} (-1)^{k-1} Bl\left(A_{i_1} \cap A_{i_1} \cap \cdots \cap A_{i_k}\right) \tag{1.16}$$

holds if and only if there exists a basic probability assignment $m\colon \mathcal{P}(X) \to [0, 1]$ *on the set* X *such that*

$$Bl(A) = \sum_{B|B\subseteq A} m(B), \tag{1.17}$$

holds for any $A \in \mathcal{P}(X)$.

Proof In Proposition 1.4, we demonstrated that if $m\colon \mathcal{P}(X) \to [0, 1]$ is a basic probability assignment on the finite set X and Eq. (1.17) holds for any $A \in \mathcal{P}(X)$; that is, the function Bl satisfies the requirement for a belief measure given in Definition 1.8, then the inequality in Eq. (1.16) holds for any $A_1, A_2, \ldots, A_n \in \mathcal{P}(X)$.

Now, we will show that if $Bl\colon \mathcal{P}(X) \to [0, 1]$ is a measure on X, $Bl(\emptyset) = 0$, $Bl(X) = 1$ and for any $A_1, A_2, \ldots, A_n \in \mathcal{P}(X)$ the function Bl satisfies the inequality in Eq. (1.16); that is, the function Bl satisfies the requirements for a belief measure given in Definition 1.4, then there exists a basic probability assignment $m\colon \mathcal{P}(X) \to [0, 1]$ on the set X such that (1.17) holds for any $A \in \mathcal{P}(X)$. Let Bl be a set function that satisfies the requirements for a belief measure given in Definition 1.4. Furthermore, let the set function $m\colon \mathcal{P}(X) \to \mathbb{R}$ be given by

$$m(A) = \sum_{B|B\subseteq A} (-1)^{|A|-|B|} Bl(B), \tag{1.18}$$

for any $A \in \mathcal{P}(X)$. Here, we will demonstrate that function m meets the criteria for a basic probability assignment given in Definition 1.7. That is,

(1) $m(\emptyset) = 0$
(2) $\sum_{A\in\mathcal{P}(X)} m(A) = 1$
(3) $0 \le m(A) \le 1$ for any $A \in \mathcal{P}(X)$.

(1) Since $Bl(\emptyset) = 0$ by definition and based on Eq. (1.18), $m(\emptyset) = Bl(\emptyset)$, $m(\emptyset) = 0$ immediately follows. (2) Applying Eq. (1.18) with $A = X$, we have

$$m(X) = \sum_{B|B\subseteq X} (-1)^{|X|-|B|} Bl(B),$$

which, based on Proposition 1.5, is equivalent to

$$Bl(X) = \sum_{B|B\subseteq X} m(B),$$

and as $Bl(X) = 1$, we get

$$1 = Bl(X) = \sum_{B|B\subseteq X} m(B) = \sum_{A\in\mathcal{P}(X)} m(A). \tag{1.19}$$

(3) Here, we will show that $m(A) \ge 0$ for any $A \in \mathcal{P}(X)$. Since A is a finite set, we can write $A = \{x_1, \ldots, x_n\}$. Let $A_i = A \setminus \{x_i\}$. Then, $A = \bigcup_{i=1}^{n} A_i$ and

$$m(A) = \sum_{B|B\subseteq A} (-1)^{|A|-|B|} Bl(B) =$$

$$= Bl(A) - \sum_{k=1}^{n} \sum_{1\leq i_1<i_2<\cdots<i_k\leq n} (-1)^{k-1} Bl\left(A_{i_1} \cap A_{i_1} \cap \cdots \cap A_{i_k}\right) =$$
$$= Bl(A_1 \cup A_2 \cup \cdots \cup A_n) -$$
$$-\sum_{k=1}^{n} \sum_{1\leq i_1<i_2<\cdots<i_k\leq n} (-1)^{k-1} Bl\left(A_{i_1} \cap A_{i_1} \cap \cdots \cap A_{i_k}\right). \tag{1.20}$$

Now, by noting Eq. (1.16), from Eq. (1.20) we get that $m(A) \geq 0$. By taking into account this result and Eq. (1.19), we also have $0 \leq m(A) \leq 1$ for any $A \in \mathcal{P}(X)$.

Next, based on Proposition 1.5, we have that for any $A \in \mathcal{P}(X)$, if Eq. (1.18) holds, then Eq. (1.17) holds as well. □

The next theorem demonstrates that the two definitions for plausibility measure given in Definitions 1.5 and 1.9 are equivalent.

Theorem 1.2 *Let X be a finite set and let $Pl\colon \mathcal{P}(X) \to [0, 1]$ be a measure on X such that $Pl(\emptyset) = 0$ and $Pl(X) = 1$. Then, for any $A_1, A_2, \ldots, A_n \in \mathcal{P}(X)$, the inequality*

$$Pl(A_1 \cap A_2 \cap \cdots \cap A_n) \leq$$
$$\geq \sum_{k=1}^{n} \sum_{1\leq i_1<i_2<\cdots<i_k\leq n} (-1)^{k-1} Pl\left(A_{i_1} \cup A_{i_1} \cup \cdots \cup A_{i_k}\right) \tag{1.21}$$

holds if and only if there exists a basic probability assignment $m\colon \mathcal{P}(X) \to [0, 1]$ on the set X such that

$$Pl(A) = \sum_{B|B\cap A\neq\emptyset} m(B), \tag{1.22}$$

holds for any $A \in \mathcal{P}(X)$.

Proof Firstly, we will show that If $m\colon \mathcal{P}(X) \to [0, 1]$ is a basic probability assignment on the finite set X and Eq. (1.22) holds for any $A \in \mathcal{P}(X)$; that is, the function Pl satisfies the requirement for a plausibility measure given in Definition 1.9, then the inequality in Eq. (1.21) holds for any $A_1, A_2, \ldots, A_n \in \mathcal{P}(X)$. Let the set function $Bl\colon \mathcal{P}(X) \to [0, 1]$ be given as $Bl(A) = 1 - Pl(\overline{A})$ for any $A \in \mathcal{P}(X)$. Then, $Pl(A) = 1 - Bl(\overline{A})$ holds as well for any $A \in \mathcal{P}(X)$. Utilizing this fact and the definition of $Pl(A)$ in Eq. (1.22), the application of Corollary 1.3 leads us to the result that Eq. (1.17) holds for any $A \in \mathcal{P}(X)$. Thus, the function Bl and the basic probability assignment m satisfy the conditions of Theorem 1.1, from which the inequality in Eq. (1.16) follows. Hence, by noting Proposition 1.3, we get that the inequality in Eq. (1.21) holds for any $A_1, A_2, \ldots, A_n \in \mathcal{P}(X)$.

Secondly, we will show that if $Pl\colon \mathcal{P}(X) \to [0, 1]$ is a measure on X, $Pl(\emptyset) = 0$, $Pl(X) = 1$ and for any $A_1, A_2, \ldots, A_n \in \mathcal{P}(X)$ the function Pl satisfies the inequality in Eq. (1.21); that is, the function Pl satisfies the requirement for a plausibility measure given in Definition 1.5, then there exists a basic probability assignment $m: \mathcal{P}(X) \to [0, 1]$ on the set X such that Eq. (1.22) holds for any $A \in \mathcal{P}(X)$. Here again, let the set function $Bl: \mathcal{P}(X) \to [0, 1]$ be given as $Bl(A) = 1 - Pl(\overline{A})$ for any $A \in \mathcal{P}(X)$. Then, based on Proposition 1.3, the function Bl satisfies the requirements for a belief measure given in Definition 1.4. Next, by applying Theorem 1.1, we get that there exists a basic probability assignment $m : \mathcal{P}(X) \to [0, 1]$ on the set X such that Eq. (1.17) holds for any $A \in \mathcal{P}(X)$. Noting that $Pl(A) = 1 - Bl(\overline{A})$ holds and applying Corollary 1.4 leads us to the result that Eq. (1.22) holds as well for any $A \in \mathcal{P}(X)$. □

Note that the belief- and plausibility measures were introduced by Dempster [1] under the names upper- and lower probabilities, induced by a probability measure by a multivalued mapping. The following propositions are about the connections among the belief-, plausibility- and probability measures.

Proposition 1.6 *If X is a finite set, Bl and Pl are belief- and plausibility measures on the set X, respectively, induced from the same basic probability assignment $m: \mathcal{P}(X) \to [0, 1]$, then for any $A \in \mathcal{P}(X)$*

$$Bl(A) \le Pl(A).$$

Proof This proposition readily follows from the definitions for belief- and plausibility measures given in Definitions 1.8 and 1.9.

Proposition 1.7 *If Pr is a probability measure on the finite set X, then Pr is both a belief- and a plausibility measure on X.*

Proof Since Pr is a probability measure on the finite set X,

$$\sum_{x \in X} Pr(\{x\}) = 1$$

holds. Let

$$m(A) = \begin{cases} Pr(A) & \text{if } |A| = 1 \\ 0 & \text{otherwise} \end{cases}$$

for any $A \in \mathcal{P}(X)$. Then, m is a basic probability assignment on the set X and

$$Pr(A) = \sum_{x \in A} Pr(\{x\}) = \sum_{B \subseteq A} m(B) = \sum_{B \cap A \neq \emptyset} m(B)$$

holds for any $A \in \mathcal{P}(X)$, which means that Pr is also a belief- and plausibility measure on the set X (Also see Remark 1.2). □

Proposition 1.8 *Let Bl and Pl be a belief- and a plausibility measure on the finite set X, respectively, induced from the same basic probability assignment m. Then Bl and Pl coincide if and only if m focuses only on the singletons of $\mathcal{P}(X)$.*

Proof Firstly, we will show that if m is a basic probability assignment on the finite set X and m focuses only on the singletons of $\mathcal{P}(X)$, then the belief- and plausibility measures induced from m coincide. Let m be a basic probability assignment on the finite set X such that $m(A) = 0$ for any $A \in \mathcal{P}(X)$, $|A| \neq 1$. Furthermore, let Bl and Pl be a belief- and a plausibility measure on X, respectively, both induced from m. Then, making use of the definitions for belief- and plausibility measures given in Definitions 1.8 and 1.9, respectively, we have

$$Bl(A) = \sum_{B \subseteq A} m(B) = \sum_{x \in A} m(\{x\}) = \sum_{B \cap A \neq \emptyset} m(B) = Pl(A)$$

for any $A \in \mathcal{P}(X)$, which means that Bl and Pl coincide.

Secondly, we will show that if Bl and Pl are belief- and a plausibility measures on X, respectively, both induced from the same basic probability assignment m and $Bl(A) = Pl(A)$ holds for any $A \in \mathcal{P}(X)$, then m focuses only on the singletons of $\mathcal{P}(X)$. Here, we will give a proof by contraposition. If there exists $A \in \mathcal{P}(X)$ such that $|A| > 1$ and $m(A) > 0$; that is, m does not focus only on the singletons of $\mathcal{P}(X)$, then for any $x \in A$

$$Bl(\{x\}) = m(\{x\}) < m(\{x\}) + m(A) \leq \sum_{B \cap \{x\} \neq \emptyset} m(B) = Pl(\{x\}),$$

which means that $Bl(\{x\})$ and $Pl(\{x\})$ do not coincide. □

As regards Propositions 1.7 and 1.8, see also Theorems 4.19 and 4.20 in [15].

Proposition 1.9 *Let Pr be a probability measure on the finite set X, and let Bl and Pl be a belief- and a plausibility measure on X, respectively, induced from the same basic probability assignment m:* $\mathcal{P}(X) \to [0, 1]$ *such that for any $x \in X$*

$$m(\{x\}) \leq Pr(\{x\}). \tag{1.23}$$

Then, the inequalities

$$Bl(A) \leq Pr(A) \leq Pl(A)$$

hold for any $A \in \mathcal{P}(X)$.

Proof Firstly, we will show that if the conditions of the proposition are satisfied, then $Bl(A) \leq Pr(A)$. If A is a singleton of $\mathcal{P}(X)$; that is, $A = \{x\}$ for some $x \in X$, then $Bl(A) = m(\{x\})$ and $Pr(A) = Pr(\{x\})$ from which $Bl(A) \leq Pr(A)$ follows. If $A \in \mathcal{P}(X)$ is not a singleton; that is, $|A| > 1$, then by noting Eq. (1.23), we have

$$\sum_{x \in A} (Pr(\{x\}) - m(\{x\})) \geq \sum_{\substack{B \subseteq A \\ |B| > 1}} m(B). \tag{1.24}$$

In this case, by utilizing Eq. (1.24) and the definition for a belief measure given by Definition 1.8, we have

$$Bl(A) = \sum_{B \subseteq A} m(B) = \sum_{\substack{B \subseteq A \\ |B|=1}} m(B) + \sum_{\substack{B \subseteq A \\ |B|>1}} m(B) =$$

$$= \sum_{x \in A} m(\{x\}) + \sum_{\substack{B \subseteq A \\ |B|>1}} m(B) \leq \sum_{x \in A} m(\{x\}) + \sum_{x \in A} (Pr(\{x\}) - m(\{x\})) =$$

$$= \sum_{x \in A} Pr(\{x\}) = Pr(A).$$

Next, by applying the result $Bl(A) \leq Pr(A)$ to the complement set $\overline{A}$, we get

$$Bl(\overline{A}) \leq Pr(\overline{A}),$$

from which we have

$$1 - Bl(\overline{A}) \geq 1 - Pr(\overline{A}).$$

Since Bl and Pl are induced from the same basic probability assignment, $1 - Bl(\overline{A}) = Pl(A)$ holds for any $A \in \mathcal{P}(X)$. Furthermore, as Pr is a probability measure, $1 - Pr(\overline{A}) = Pr(A)$ also holds for any $A \in \mathcal{P}(X)$. Therefore, the previous inequality can be written as

$$Pl(A) \geq 1 - Pr(\overline{A}) = Pr(A).$$

That is, we get the result that the inequality $Bl(A) \leq Pr(A) \leq Pl(A)$ holds for any $A \in \mathcal{P}(X)$. □

Proposition 1.9 allows us to estimate unknown probabilities by a dual pair of belief- and plausibility measures when a basic probability assignment is given. Namely, viewing values $Bl(A)$ and $Pl(A)$ as lower- and upper probabilities, we have the interval $[Bl(A), Pl(A)]$ for the estimated probability $Pr(A)$.

1.3.2 Examples

Example 1.1 Suppose that there are three securities, "a", "b" and "c", which we may invest in. We know that there is only one among them that will have the highest yield. Our broker states that "*a*" or "*b*" will have the highest yield among the three securities. We also know that our broker gives good advice in 80% of all the cases. Let a, b and c denote the elementary events that the security with the highest yield is "a", "b" and "c", respectively. Then, utilizing the concept of basic probability assignment, the available information can be represented by $m(\{a, b\}) = 0.8$ and

Table 1.1 The computed belief- and plausibility measure values

$A \in \mathcal{P}(X)$	$\emptyset$	$\{a\}$	$\{b\}$	$\{c\}$	$\{a, b\}$	$\{a, c\}$	$\{b, c\}$	X
$m(A)$	0	0	0	0	0.8	0	0	0.2
$Bl(A)$	0	0	0	0	0.8	0	0	1
$Pl(A)$	0	1	1	0.2	1	1	1	1

$m(X) = 0.2$, where $X = \{a, b, c\}$. Notice that the mass 0.2 is not committed to c since there is no information (evidence) that supports it! Table 1.1 summarizes the computed belief- and plausibility measure values for each subset of X.

Based on the data in Table 1.1, we can state that the belief in the event $\{a, b\}$; that is, in the event that the securities "a" or "b" will have the highest yield has a value of 0.8, while the same event is plausible with the value of 1.

We can also see that the belief in event $\{c\}$ is zero, and at the same time this event is plausible with the value of 0.2. It means that even we have no belief in event $\{c\}$, we still find it (slightly) plausible.

Here, the complement $\overline{E}$ of event $E = \{c\}$ is $\overline{E} = \{a, b\}$. Notice that $Bl(E) = 0$ and $Bl(\overline{E}) = 0.8$; that is, lack of belief in E does not imply a belief measure value of 1 for the event $\overline{E}$. Notice that this finding is in accordance with the result of Lemma 1.1; namely, $Bl(E) + Bl(\overline{E}) \leq 1$. Also, $Pl(E) = 0.2$ and $Pl(\overline{E}) = 1$, which means that, in accordance with Lemma 1.2, the sum of plausibility of an event and its complement may be greater than or equal to one; that is, $Pl(E) + Pl(\overline{E}) \geq 1$.

We can see from Table 1.1 that the measures Bl and Pl are not additive. We can also notice the main properties of these measures. For example,

$$Bl(\{a, b\} \cup \{a, c\}) = Bl(\{a, b, c\}) = 1$$
$$Bl(\{a, b\}) + Bl(\{a, c\}) - Bl(\{a, b\} \cap \{a, c\}) = 0.8,$$

that is,

$$Bl(\{a, b\} \cup \{a, c\}) \geq Bl(\{a, b\}) + Bl(\{a, c\}) - Bl(\{a, b\} \cap \{a, c\}).$$

Also,

$$Pl(\{a, c\} \cap \{b, c\}) = Pl(\{c\}) = 0.2$$
$$Pl(\{a, c\}) + Pl(\{b, c\}) - Pl(\{a, c\} \cup \{b, c\}) = 1,$$

which means that

$$Pl(\{a, c\} \cap \{b, c\}) \leq Pl(\{a, c\}) + Pl(\{b, c\}) - Pl(\{a, c\} \cup \{b, c\}).$$

Example 1.2 (*Ellsberg's Paradox* [16]) Let us assume that we have an urn which contains 30 red balls and 60 other balls that are either black or yellow. It is not known how many black and how many yellow balls are in the urn; we only know that the sum of the total number of black balls and the total number of yellow balls is 60. Each ball can be drawn from the urn with the same likelihood.

You have a choice between the following bets:

- A: You receive 100 euros, if you draw a red ball
- B: You receive 100 euros, if you draw a black ball

You also have a choice between these two bets (another draw from the same urn):

- A': You receive 100 euros, if you draw a red or yellow ball
- B': You receive 100 euros, if you draw a black or yellow ball

Which bet would you prefer in the first case, and which one in the second case? Most people strictly prefer bet A to bet B in the first case, and bet B' to bet A' in the second case. Formally,

$$A \succ B \text{ and } B' \succ A'.$$

Since the outcome of a draw is either a red or a black or a yellow ball, the event space is $X = \{r, b, y\}$, where r, b and y denote the red, black and yellow balls, respectively. Let us assume that according to the expected utility theory [17], the preferences are based on a subjective probability Pr, which is additive. That is, bet A is preferred to bet B, if and only if

$$Pr(\{r\}) > Pr(\{b\}), \tag{1.25}$$

and bet B' is preferred to bet A', if and only if

$$Pr(\{b, y\}) > Pr(\{r, y\}).$$

By noting the fact that Pr is an additive measure, from the last inequality we get

$$Pr(\{b\}) + Pr(\{y\}) > Pr(\{r\}) + Pr(\{y\}),$$

from which

$$Pr(\{b\}) > Pr(\{r\}) \tag{1.26}$$

follows. Notice that Eq. (1.25) contradicts Eq. (1.26). This result means that the application of additive measures is not the best approach to model preferences if only partial knowledge is available concerning the probability of some events. In this example, the exact probabilities of the events related to bet A and bet B' are known; these are

$$Pr(\{r\}) = \frac{1}{3}$$

and

$$Pr(\{b, y\}) = \frac{2}{3},$$

Table 1.2 The computed belief- and plausibility measure values

$E \in \mathcal{P}(X)$	$\emptyset$	$\{r\}$	$\{b\}$	$\{y\}$	$\{r, b\}$	$\{r, y\}$	$\{b, y\}$	X
$m(E)$	0	1/3	0	0	0	0	2/3	0
$Bl(E)$	0	1/3	0	0	1/3	1/3	2/3	1
$Pl(E)$	0	1/3	2/3	2/3	1	1	2/3	1

respectively. At the same time, the exact probabilities of the events related to bet B and bet A' are unknown; we only have partial knowledge concerning them. Namely, we know that

$$0 \leq Pr(\{b\}) \leq \frac{2}{3}$$

and

$$\frac{1}{3} \leq Pr(\{r, y\}) \leq 1.$$

Modeling with Belief- and Plausibility Measures

Let the function $m \colon \mathcal{P}(X) \to [0, 1]$ be a basic probability assignment function. Then, the available information can be represented by the mass values

$$m(\{r\}) = \frac{1}{3}, m(\{b, y\}) = \frac{2}{3}$$

and $\forall E \in \mathcal{P}(X)$, $E \neq \{r\}$, $E \neq \{y, b\}$: $m(E) = 0$. Table 1.2 summarizes the computed belief- and plausibility measure values for each subset of X.

Here, the belief measure values $Bl(\{r\}) = 1/3$, $Bl(\{b\}) = 0$, $Bl(\{r, y\}) = 1/3$ and $Bl(\{b, y\}) = 2/3$ can be interpreted as the degrees to which the available information (evidence) supports the events $\{r\}$, $\{b\}$, $\{r, y\}$ and $\{b, y\}$, respectively. Therefore, if we base our preference on the belief measure values of the events related to the bets, then, since

$$\underbrace{Bl(\{r\})}_{1/3} > \underbrace{Bl(\{b\})}_{0} \tag{1.27}$$

and

$$\underbrace{Bl(\{r, y\})}_{1/3} < \underbrace{Bl(\{b, y\})}_{2/3} \tag{1.28}$$

hold, we prefer bet A to bet B, and prefer bet B' to bet A'. Notice that according to the properties of the belief measure, the inequalities

$$\underbrace{Bl(\{r, y\})}_{1/3} \geq \underbrace{Bl(\{r\})}_{1/3} + \underbrace{Bl(\{y\})}_{0}$$

and

$$\underbrace{Bl(\{b,y\})}_{2/3} \geq \underbrace{Bl(\{b\})}_{0} + \underbrace{Bl(\{y\})}_{0}$$

hold and these inequalities do not contradict the inequalities in Eqs. (1.27) and (1.28).

The plausibility values $Pl(\{r\}) = 1/3$, $Pl(\{b\}) = 2/3$, $Pl(\{r, y\}) = 1$ and $Pl(\{b, y\}) = 2/3$ can be interpreted as the upper boundaries on the degrees to which the available information could support the events $\{r\}$, $\{b\}$, $\{r, y\}$ and $\{b, y\}$, respectively. Now, supposing that the preferences of the bets are based on the plausibility measure values of the events related to them, since

$$\underbrace{Pl(\{r\})}_{1/3} < \underbrace{Pl(\{b\})}_{2/3} \tag{1.29}$$

and

$$\underbrace{Pl(\{r,y\})}_{1} > \underbrace{Pl(\{b,y\})}_{2/3} \tag{1.30}$$

hold, we prefer bet B to bet A, and prefer bet A' to bet B'. We can also see that, in accordance with the properties of the plausibility measure, the inequalities

$$\underbrace{Pl(\{r,y\})}_{1} \leq \underbrace{Pl(\{r\})}_{1/3} + \underbrace{Pl(\{y\})}_{2/3}$$

and

$$\underbrace{Pl(\{b,y\})}_{2/3} \leq \underbrace{Pl(\{b\})}_{2/3} + \underbrace{Bl(\{y\})}_{2/3}$$

hold and these inequalities do not contradict the inequality in Eqs. (1.29) and (1.30).

Notice that the belief- and plausibility measures may be viewed as lower- and upper probabilities. Here, for the unknown probabilities $Pr(\{b\})$ and $Pr(\{r, y\})$ we have

$$\underbrace{Bl(\{b\})}_{0} \leq Pr(\{b\}) \leq \underbrace{Pl(\{b\})}_{2/3}$$

and

$$\underbrace{Bl(\{r,y\})}_{1/3} \leq Pr(\{r,y\}) \leq \underbrace{Pl(\{r,y\})}_{1}.$$

That is, we do not know the exact probability of drawing a black ball, but we consider this event being plausible at 2/3. Similarly, the exact probability of drawing a red or yellow ball is unknown, but based on the available information, this event has a belief value of 1/3.

Example 1.3 Consider two suppliers, SUP1 and SUP2, who supply a component to a company. We know that S1 is a qualified suppler and always delivers good components to the company. We do not have any information about the quality of components supplied by supplier SUP2. Suppose that the production uses components from supplier SUP1 with a probability of 0.8 and from supplier SUP2 with a probability of 0.2. What is the likelihood of the event that a bad component is built into a product? To answer this question, we will formalize the problem as follows. The outcome of a component selection is a good or a bad component, so the event space is $X = \{g, b\}$, where g and b denote the good and bad components, respectively. Furthermore, let S_1 and S_2 denote the events that the selected component is from supplier SUP1 and SUP2, respectively. Here, we know the probability values

$$Pr(S_1) = 0.8,\ Pr(S_2) = 0.2,$$

and the conditional probability values

$$Pr(\{g\}|S_1) = 1,\ Pr(\{b\}|S_1) = 0,$$

but we do not know the exact values of the probabilities $Pr(\{g\}|S_2)$ and $Pr(\{b\}|S_2)$. We only know that

$$Pr(\{g\}|S_2) + Pr(\{b\}|S_2) = 1.$$

Then, the available information can be represented by the mass values

$$m(\{g\}) = 0.8,\ m(\{g, b\}) = 0.2$$

and $\forall E \in \mathcal{P}(X),\ E \neq \{g\},\ E \neq \{g, b\}\colon m(E) = 0$, where the function $m\colon \mathcal{P}(X) \to [0, 1]$ is a basic probability assignment function. Table 1.3 summarizes the computed belief- and plausibility measure values for each subset of X. By having the values of $Bl\{g\}$, $Pl\{g\}$, $Bl\{b\}$ and $Pl\{b\}$, the probabilities $Pr\{g\}$ and $Pr\{b\}$ can be estimated as

$$\underbrace{Bl(\{g\})}_{0.8} \leq Pr(\{g\}) \leq \underbrace{Pl(\{g\})}_{1}$$

$$\underbrace{Bl(\{b\})}_{0} \leq Pr(\{b\}) \leq \underbrace{Pl(\{b\})}_{0.2}.$$

Table 1.3 The computed belief- and plausibility measure values

$E \in \mathcal{P}(X)$	$\emptyset$	$\{g\}$	$\{b\}$	$\{g, b\}$
$m(E)$	0	0.8	0	0.2
$Bl(E)$	0	0.8	0	1
$Pl(E)$	0	1	0.2	1

That is, the likelihood of the event that a bad component will be built into a product is between 0 and 0.2. Notice that the same estimates can be obtained from a probabilistic approach as well. Namely,

$$Pr(\{g\}) = \underbrace{Pr(\{g\}|S_1)}_{1}\underbrace{Pr(S_1)}_{0.8} + \underbrace{Pr(\{g\}|S_2)}_{\in[0,1]}\underbrace{Pr(S_2)}_{0.2},$$

and so

$$0.8 \leq Pr(\{g\}) \leq 1.$$

Similarly,

$$Pr(\{b\}) = \underbrace{Pr(\{b\}|S_1)}_{0}\underbrace{Pr(S_1)}_{0.8} + \underbrace{Pr(\{b\}|S_2)}_{\in[0,1]}\underbrace{Pr(S_2)}_{0.2},$$

and so

$$0 \leq Pr(\{b\}) \leq 0.2.$$

We should also add that the approach based on the belief- and plausibility measures gives us the lower- and upper estimates of the probabilities in a way that is simpler than the probabilistic approach.

1.4 Summary

In this chapter, we briefly discussed three important measures: the belief-, the probability- and the plausibility measures as these play important roles in describing various phenomena. We presented a short summary of the main properties of these measures and outlined their applicability in modeling uncertainty. Here, by following Dempster–Shafer's theory of evidence, we outlined how the belief- and plausibility measures can be utilized in situations where the probabilistic approach is not applicable or not suitable for representing information. By the means of examples, we showed how the belief- and plausibility measures, as lower- and upper probabilities can be used in practice.

References

1. A.P. Dempster, Upper and lower probabilities induced by a multivalued mapping. Ann. Math. Stat. **38**, 325–339 (1967)
2. G. Shafer, *A Mathematical Theory of Evidence* (Princeton University Press, Princeton, 1976)
3. N. Friedman, J.Y. Halpern, Plausibility measures and default reasoning. J. ACM **48**(4), 648–685 (2001)
4. D. Dubois, H. Prade, *Fuzzy Sets and Systems: Theory and Applications*. Mathematics in Science and Engineering, vol. 144 (Academic, Orlando, 1980)

5. T.L. Fine, *Theories of Probability: an Examination of Foundations* (Academic, New York, 2014)
6. O. Kanjanatarakul, T. Denœux, S. Sriboonchitta, Prediction of future observations using belief functions: a likelihood-based approach. Int. J. Approx. Reason. **72**, 71–94 (2016). https://doi.org/10.1016/j.ijar.2015.12.004
7. R. Jiroušek, P.P. Shenoy, A new definition of entropy of belief functions in the Dempster-Shafer theory. Int. J. Approx. Reason. **92**, 49–65 (2018). https://doi.org/10.1016/j.ijar.2017.10.010
8. L. Jiao, Q. Pan, T. Denœux, Y. Liang, X. Feng, Belief rule-based classification system: extension of FRBCS in belief functions framework. Inf. Sci. **309**, 26–49 (2015). https://doi.org/10.1016/j.ins.2015.03.005
9. Y. Liu, Q. Cheng, Q. Xia, X. Wang, The use of evidential belief functions for mineral potential mapping in the Nanling belt, South China. Front. Earth Sci. **9**(2), 342–354 (2015). https://doi.org/10.1007/s11707-014-0465-4
10. D. Han, J. Dezert, Z. Duan, Evaluation of probability transformations of belief functions for decision making. IEEE Trans. Syst. Man Cybern.: Syst. **46**(1), 93–108 (2016). https://doi.org/10.1109/TSMC.2015.2421883
11. J. Dombi, T. Jónás, The λ-additive measure in a new light: the Q_ν measure and its connections with belief, probability, plausibility, rough sets, multi-attribute utility functions and fuzzy operators. Soft Comput. (2019). https://doi.org/10.1007/s00500-019-04212-y
12. J. Dombi, T. Jónás, The ν-additive measure as an alternative to the λ-additive measure, in *38th Linz Seminar on Fuzzy Set Theory*, ed. by M. Grabish, T. Kroupa, S. Saminger-Platz, T. Vetterlein (2019), pp. 26–29
13. G. Choquet, Theory of capacities. Annales de l'institut Fourier **5**, 131–295 (1954)
14. M. Sugeno, Theory of fuzzy integrals and its applications. Ph.D. thesis, Tokyo Institute of Technology, Tokyo, Japan (1974)
15. Z. Wang, G.J. Klir, *Generalized Measure Theory*. IFSR International Series in Systems Science and Systems Engineering (Springer US, New York, 2010)
16. D. Ellsberg, Risk, ambiguity, and the savage axioms. Q. J. Econ. **75**(4), 643–669 (1961)
17. L.J. Savage, *The Foundations of Statistics*. Wiley Publications in Statistics (1954)

Chapter 2
λ-Additive and ν-Additive Measures

In many situations, the application of traditional additive measures is not sufficient to describe the uncertainty appropriately. Therefore, new demands have arisen for not necessarily additive, but monotone (fuzzy) measures. Since these measures play an important role in describing various phenomena, there is an increasing interest in them (see, e.g. [1–5]). Undoubtedly, one of the most widely applied classes of monotone measures is the class of λ-additive measures (Sugeno λ-measures) [6]. In this chapter, we will discuss the key properties of λ-additive measures, introduce an alternatively parameterized λ-additive measure, which we will call the ν-additive measure, and demonstrate our novel results that are related to these two measures. Furthermore, it will also be shown here how the ν-additive (λ-additive) measures can be utilized for characterizing 'plausibilitiness' or 'beliefness' as well as how these two measures are connected with rough sets, multi-attribute utility functions and with certain operators of fuzzy logic. This chapter is connected with the following publications of ours: [7–12].

2.1 The λ-Additive Measure

Relaxing the additivity property of the probability measure, the λ-additive measures were proposed by Sugeno in 1974 [6].

Definition 2.1 The function $Q_\lambda \colon \mathcal{P}(X) \to [0, 1]$ is a λ-additive measure (Sugeno λ-measure) on the finite set X, if and only if Q_λ satisfies the following requirements:

(1) $Q_\lambda(X) = 1$
(2) for any $A, B \in \mathcal{P}(X)$ and $A \cap B = \emptyset$,

J. Dombi and T. Jónás, *Advances in the Theory of Probabilistic and Fuzzy Data Scientific Methods with Applications*, Studies in Computational Intelligence 814,
https://doi.org/10.1007/978-3-030-51949-0_2

$$Q_\lambda(A \cup B) = Q_\lambda(A) + Q_\lambda(B) + \lambda Q_\lambda(A) Q_\lambda(B), \tag{2.1}$$

where $\lambda \in (-1, \infty)$ and $\mathcal{P}(X)$ is the power set of X.

2.1.1 Basic Properties of the λ-Additive Measure

Now, we will show that any λ-additive measure is a monotone (fuzzy) measure.

Proposition 2.1 *Every λ-additive measure is a monotone (fuzzy) measure.*

Proof Let Q_λ be a λ-additive measure on the set X. Then $Q_\lambda(X) = 1$ holds by definition. Next, by utilizing Eq. (2.1), we get $Q_\lambda(X) = Q_\lambda(X \cup \emptyset) = Q_\lambda(X) + Q_\lambda(\emptyset)(1 + \lambda Q_\lambda(X))$, which implies that $Q_\lambda(\emptyset) = 0$. Thus, Q_λ satisfies criterion (1) of a monotone measure given in Definition 1.2.

Next, let $A, B \in \mathcal{P}(X)$ and let $B \subseteq A$. Then there exists a $C \in \mathcal{P}(X)$ such that $A = B \cup C$ and $B \cap C = \emptyset$. Now, by utilizing Eq. (2.1) and the fact that $\lambda > -1$, we get

$$Q_\lambda(A) = Q_\lambda(B \cup C) =$$
$$= Q_\lambda(B) + Q_\lambda(C)(1 + \lambda Q_\lambda(B)) \geq Q_\lambda(B).$$

It means that Q_λ also satisfies the monotonicity criterion of a monotone (fuzzy) measure. □

Here, we will show how the λ-additive measure of difference of two sets can be computed.

Proposition 2.2 *If X is a finite set and Q_λ is a λ-additive measure on X, then for any $A, B \in \mathcal{P}(X)$ the Q_λ measure of the set difference $A \setminus B$ is*

$$Q_\lambda(A \setminus B) = \frac{Q_\lambda(A) - Q_\lambda(A \cap B)}{1 + \lambda Q_\lambda(A \cap B)} \tag{2.2}$$

(see Theorem 4.6 (1) in [3]).

Proof Since $(A \cap B) \cup (A \setminus B) = A$ and $(A \cap B) \cap (A \setminus B) = \emptyset$, we can write

$$Q_\lambda(A) = Q_\lambda\left((A \cap B) \cup (A \setminus B)\right) =$$
$$= Q_\lambda(A \cap B) + Q_\lambda(A \setminus B) + \lambda Q_\lambda(A \cap B) Q_\lambda(A \setminus B) =$$
$$= Q_\lambda(A \cap B) + Q_\lambda(A \setminus B)(1 + \lambda Q_\lambda(A \setminus B)),$$

from which we get

$$Q_\lambda(A \setminus B) = \frac{Q_\lambda(A) - Q_\lambda(A \cap B)}{1 + \lambda Q_\lambda(A \cap B)}.$$

□

Here, we will show how the λ-additive measure of a complement set can be computed.

Proposition 2.3 *If X is a finite set and Q_λ is a λ-additive measure on X, then for any $A \in \mathcal{P}(X)$ the Q_λ measure of the complement set $\overline{A} = X \setminus A$ is*

$$Q_\lambda(\overline{A}) = \frac{1 - Q_\lambda(A)}{1 + \lambda Q_\lambda(A)} \tag{2.3}$$

(see Theorem 4.6 (3) in [3]).

Proof Since $\overline{A} = X \setminus A$, using Proposition 2.2, we have

$$Q_\lambda(\overline{A}) = Q_\lambda(X \setminus A) = \frac{Q_\lambda(X) - Q_\lambda(X \cap A)}{1 + \lambda Q_\lambda(X \cap A)} = \frac{1 - Q_\lambda(A)}{1 + \lambda Q_\lambda(A)}.$$

□

The calculation of the λ-additive measure of two disjoint sets is given in Definition 2.1. Here, we will show how the λ-additive measure of two sets can be computed when these sets are not necessarily disjoint.

Proposition 2.4 *If X is a finite set and Q_λ is a λ-additive measure on X, then for any $A, B \in \mathcal{P}(X)$,*

$$Q_\lambda(A \cup B) = \frac{Q_\lambda(A) + Q_\lambda(B) + \lambda Q_\lambda(A) Q_\lambda(B) - Q_\lambda(A \cap B)}{1 + \lambda Q_\lambda(A \cap B)}$$

(see Theorem 4.6 (2) in [3]).

Proof Since $A \cap (\overline{A} \cap B) = \emptyset$ and $A \cup (\overline{A} \cap B) = A \cup B$, applying Eq. (2.1) gives us

$$\begin{aligned} Q_\lambda(A \cup B) &= Q_\lambda(A) + Q_\lambda(\overline{A} \cap B) + \lambda Q_\lambda(A) Q_\lambda(\overline{A} \cap B) = \\ &= Q_\lambda(A) + Q_\lambda(\overline{A} \cap B)(1 + \lambda Q_\lambda(A)). \end{aligned} \tag{2.4}$$

Next, since $(A \cap B) \cap (\overline{A} \cap B) = \emptyset$ and $(A \cap B) \cup (\overline{A} \cap B) = B$, applying Eq. (2.1) again gives

$$\begin{aligned} Q_\lambda(B) &= Q_\lambda(A \cap B) + Q_\lambda(\overline{A} \cap B) + \lambda Q_\lambda(A \cap B) Q_\lambda(\overline{A} \cap B) = \\ &= Q_\lambda(A \cap B) + Q_\lambda(\overline{A} \cap B)(1 + \lambda Q_\lambda(A \cap B)). \end{aligned} \tag{2.5}$$

Now, by expressing $Q_\lambda(\overline{A} \cap B)$ in terms of (2.5), we get

$$Q_\lambda(\overline{A} \cap B) = \frac{Q_\lambda(B) - Q_\lambda(A \cap B)}{1 + \lambda Q_\lambda(A \cap B)}$$

and substituting this into (2.4), we get

$$\begin{aligned} Q_\lambda(A \cup B) &= Q_\lambda(A) + \frac{Q_\lambda(B) - Q_\lambda(A \cap B)}{1 + \lambda Q_\lambda(A \cap B)}(1 + \lambda Q_\lambda(A)) = \\ &= \frac{Q_\lambda(A) + Q_\lambda(B) + \lambda Q_\lambda(A) Q_\lambda(B) - Q_\lambda(A \cap B)}{1 + \lambda Q_\lambda(A \cap B)}. \end{aligned} \tag{2.6}$$

Hence, we have the general form of the λ-additive measure of the union of two sets. □

Remark 2.1 Notice that if $\lambda = 0$, then Eq. (2.6) reduces to $Q_\lambda(A \cup B) = Q_\lambda(A) + Q_\lambda(B) - Q_\lambda(A \cap B)$, which has the same form as the probability measure of union of two sets.

The following lemma concerns the λ-additive measure of intersection of two sets.

Lemma 2.1 *If X is a finite set and Q_λ is a λ-additive measure on X, then for any $A, B \in \mathcal{P}(X)$,*

$$\begin{aligned} &Q_\lambda(A \cap B) = \\ &= \frac{Q_\lambda(A) + Q_\lambda(B) + \lambda Q_\lambda(A) Q_\lambda(B) - Q_\lambda(A \cup B)}{1 + \lambda Q_\lambda(A \cup B)}. \end{aligned} \tag{2.7}$$

Proof By expressing $Q_\lambda(A \cap B)$ in Eq. (2.6), we get Eq. (2.7). □

2.1.1.1 The General Poincaré Formula

Although there are many theoretical and practical articles (see, e.g. [13–17]) that discuss the λ-additive measure, its properties and its applicability, there are no studies that deal with the general form of the λ-additive measure of the union of n sets. In this section, we will prove that if X is a finite set, $A_1, \ldots, A_n \in \mathcal{P}(X)$, $n \geq 2$, Q_λ is a λ-additive measure on X and $\lambda \neq 0$, then

$$\begin{aligned} &Q_\lambda\left(\bigcup_{i=1}^{n} A_i\right) = \\ &= \frac{1}{\lambda}\left(\prod_{k=1}^{n}\left(\prod_{1 \leq i_1 < \cdots < i_k \leq n} \left(1 + \lambda Q_\lambda\left(A_{i_1} \cap \cdots \cap A_{i_k}\right)\right)\right)^{(-1)^{k-1}} - 1\right). \end{aligned} \tag{2.8}$$

The well-known Poincaré formula of probability theory is

$$Pr\left(\bigcup_{i=1}^{n} A_i\right) = \sum_{k=1}^{n} (-1)^{k-1} \sum_{1 \leq i_1 < \cdots < i_k \leq n} Pr\left(A_{i_1} \cap \cdots \cap A_{i_k}\right), \tag{2.9}$$

where Pr is a probability measure on X and $A_1, \ldots, A_n \in \mathcal{P}(X)$. Here, we will demonstrate that the Poincaré formula of probability theory given in Eq. (2.9) may be viewed as a special case of the general formula of λ-additive measure of union of n sets given in Eq. (2.8). Namely, we will prove that if X is a finite set, $A_1, \ldots, A_n \in \mathcal{P}(X)$, $n \geq 2$, Q_λ is a λ-additive measure on X and $\lambda \neq 0$, then

$$\lim_{\lambda \to 0} Q_\lambda \left(\bigcup_{i=1}^{n} A_i \right) = \sum_{k=1}^{n} (-1)^{k-1} \sum_{1 \leq i_1 < \cdots < i_k \leq n} Q_\lambda \left(A_{i_1} \cap \cdots \cap A_{i_k} \right).$$

Here, we will introduce a function and some quantities that we will utilize later on.

Definition 2.2 The function $p_{n,\lambda}^{(k)} \colon \mathcal{P}^n(X) \to \mathbb{R}$ is given by

$$p_{n,\lambda}^{(k)}(A_1, \ldots, A_n) = \prod_{1 \leq i_1 < \cdots < i_k \leq n} \left(1 + \lambda Q_\lambda \left(A_{i_1} \cap \cdots \cap A_{i_k}\right)\right),$$

where X is a finite set, $A_1, \ldots, A_n \in \mathcal{P}(X)$, $n \geq 2$, $1 \leq k \leq n$. For the sake of simplicity, we will also use the $Z_{n,\lambda}^{(k)} = p_{n,\lambda}^{(k)}(A_1, \ldots, A_n)$ notation.

Later, we will utilize the following quantity to identify the general formula for the λ-additive measure of union of n sets.

Definition 2.3 The quantity $Z_{n,\lambda}^{*(k)}$ is given by

$$Z_{n,\lambda}^{*(k)} = p_{n,\lambda}^{(k)}(A_1^*, \ldots, A_n^*),$$

where X is a finite set, $A_i^* = A_i \cap A_{n+1}$, $A_i, A_{n+1} \in \mathcal{P}(X)$, $n \geq 2$, $1 \leq i \leq n$, $1 \leq k \leq n$.

Here, we will demonstrate how the λ-additive measure of union of n general sets can be computed. That is, we will discuss the Poincaré formula for the λ-additive measure. First, we will discuss some key properties of the quantities that we introduced previously.

Lemma 2.2 *If X is a finite set, $A_1, \ldots, A_n, A_{n+1} \in \mathcal{P}(X)$, $A_i^* = A_i \cap A_{n+1}$ and $1 \leq i \leq n$, then*

$$Z_{n,\lambda}^{*(k)} = p_{n,\lambda}^{(k)}(A_1^*, \ldots, A_n^*) = \\ = \prod_{1 \leq i_1 < \cdots < i_k \leq n} \left(1 + \lambda Q_\lambda \left(A_{i_1} \cap \cdots \cap A_{i_k} \cap A_{n+1}\right)\right)$$

where $n \geq 2$ and $1 \leq k \leq n$.

Proof Exploiting the idempotent property of the set intersection operation, the lemma immediately follows from the definition of $Z_{n,\lambda}^{*(k)}$. □

The following lemma demonstrates a key connection between the $Z^{*(k)}_{n,\lambda}$ and $Z^{(n)}_{n,\lambda}$ quantities.

Lemma 2.3 *Let X be a finite set and let $A_1, \ldots, A_n, A_{n+1} \in \mathcal{P}(X)$, $A_i^* = A_i \cap A_{n+1}$ and $1 \le i \le n$. Then, for any $n \ge 2$, $1 \le k \le n$, the quantity $Z^{*(k)}_{n,\lambda}$ can be expressed in terms of $Z^{(k+1)}_{n,\lambda}$ and $Z^{(k+1)}_{n+1,\lambda}$ as follows:*

$$Z^{*(k)}_{n,\lambda} = \begin{cases} \frac{Z^{(k+1)}_{n+1,\lambda}}{Z^{(k+1)}_{n,\lambda}}, & \text{if } k < n \\ Z^{(n+1)}_{n+1,\lambda}, & \text{if } k = n. \end{cases} \tag{2.10}$$

Proof Here, we will distinguish two cases: (1) $k < n$, (2) $k = n$.

(1) Based on Lemma 2.2, the following equation holds:

$$\begin{aligned} Z^{*(k)}_{n,\lambda} = p^{(k)}_{n,\lambda}(A_1^*, \ldots, A_n^*) = \\ = \prod_{1 \le i_1 < \cdots < i_k \le n} \left(1 + \lambda Q_\lambda \left(A_{i_1} \cap \cdots \cap A_{i_k} \cap A_{n+1}\right)\right). \end{aligned} \tag{2.11}$$

Next, the right hand side of Eq. (2.11) can be written as

$$\begin{aligned} \prod_{1 \le i_1 < \cdots < i_k \le n} \left(1 + \lambda Q_\lambda \left(A_{i_1} \cap \cdots \cap A_{i_k} \cap A_{n+1}\right)\right) = \\ = \frac{\prod_{1 \le i_1 < \cdots < i_{k+1} \le n+1} \left(1 + \lambda Q_\lambda \left(A_{i_1} \cap \cdots \cap A_{i_{k+1}}\right)\right)}{\prod_{1 \le i_1 < \cdots < i_{k+1} \le n} \left(1 + \lambda Q_\lambda \left(A_{i_1} \cap \cdots \cap A_{i_{k+1}}\right)\right)} = \frac{Z^{(k+1)}_{n+1,\lambda}}{Z^{(k+1)}_{n,\lambda}}. \end{aligned} \tag{2.12}$$

Notice that based on Definition 2.2, $Z^{(k+1)}_{n,\lambda}$ exists only if $k + 1 \le n$; that is, if $k < n$. This explains why we need to differentiate the two cases in Eq. (2.10).

(2) If $k = n$, then based on Definitions 2.3 and 2.2,

$$\begin{aligned} Z^{*(k)}_{n,\lambda} = p^{(k)}_{n,\lambda}(A_1^*, \ldots, A_n^*) = 1 + \lambda Q_\lambda \left(A_1^* \cap \cdots \cap A_n^*\right) = \\ = 1 + \lambda Q_\lambda \left((A_1 \cap A_{n+1}) \cap \cdots \cap (A_n \cap A_{n+1})\right) = \\ = 1 + \lambda Q_\lambda \left(A_1 \cap \cdots \cap A_n \cap A_{n+1}\right) = p^{(n+1)}_{n+1,\lambda}(A_1, \ldots, A_{n+1}) = Z^{(n+1)}_{n+1,\lambda}. \end{aligned} \tag{2.13}$$

□

The following example demonstrates the usefulness of Lemma 2.3. In this example, we will show how the quantity $Z^{*(1)}_{3,\lambda}$ can be expressed in terms of the quantities $Z^{(2)}_{4,\lambda}$ and $Z^{(2)}_{3,\lambda}$.

Example 2.1

$$Z^{*(1)}_{3,\lambda} = \left(1 + \lambda Q_\lambda \left(A_1 \cap A_4\right)\right)\left(1 + \lambda Q_\lambda \left(A_2 \cap A_4\right)\right)\left(1 + \lambda Q_\lambda \left(A_3 \cap A_4\right)\right)$$

$$Z_{4,\lambda}^{(2)} = (1+\lambda Q_\lambda(A_1 \cap A_2))(1+\lambda Q_\lambda(A_1 \cap A_3))(1+\lambda Q_\lambda(A_1 \cap A_4)) \cdot$$
$$\cdot (1+\lambda Q_\lambda(A_2 \cap A_3))(1+\lambda Q_\lambda(A_2 \cap A_4))(1+\lambda Q_\lambda(A_3 \cap A_4))$$

$$Z_{3,\lambda}^{(2)} = (1+\lambda Q_\lambda(A_1 \cap A_2))(1+\lambda Q_\lambda(A_1 \cap A_3))(1+\lambda Q_\lambda(A_2 \cap A_3))$$

It can be seen from the expressions of $Z_{3,\lambda}^{*(1)}$, $Z_{4,\lambda}^{(2)}$ and $Z_{3,\lambda}^{(2)}$ that the equation

$$Z_{3,\lambda}^{*(1)} = \frac{Z_{4,\lambda}^{(2)}}{Z_{3,\lambda}^{(2)}}$$

holds.

The next lemma shows how the λ-additive measure of set A_n can be expressed in terms of the $Z_{n,\lambda}^{(1)}$ and $Z_{n-1,\lambda}^{(1)}$ quantities.

Lemma 2.4 *If X is a finite set, $A_1, \dots, A_n \in \mathcal{P}(X)$, $n \geq 3$ and $\lambda \neq 0$, then*

$$Q_\lambda(A_n) = \frac{1}{\lambda}\left(\frac{Z_{n,\lambda}^{(1)}}{Z_{n-1,\lambda}^{(1)}} - 1\right). \tag{2.14}$$

Proof By utilizing the definitions of $Z_{n,\lambda}^{(1)}$ and $Z_{n-1,\lambda}^{(1)}$, we have

$$\frac{Z_{n,\lambda}^{(1)}}{Z_{n-1,\lambda}^{(1)}} = \frac{(1+\lambda Q_\lambda(A_1))\cdots(1+\lambda Q_\lambda(A_{n-1}))(1+\lambda Q_\lambda(A_n))}{(1+\lambda Q_\lambda(A_1))\cdots(1+\lambda Q_\lambda(A_{n-1}))} =$$
$$= 1+\lambda Q_\lambda(A_n),$$

from which Eq. (2.14) immediately follows. □

Now, we will state and prove a key theorem that allows us to compute the λ-additive measure of the union of n sets when the parameter λ is nonzero.

Theorem 2.1 *If X is a finite set, $A_1, \dots, A_n \in \mathcal{P}(X)$, $n \geq 2$, Q_λ is a λ-additive measure on X and $\lambda \neq 0$, then*

$$Q_\lambda\left(\bigcup_{i=1}^{n} A_i\right) =$$
$$= \frac{1}{\lambda}\left(\prod_{k=1}^{n}\left(\prod_{1 \leq i_1 < \cdots < i_k \leq n}(1+\lambda Q_\lambda(A_{i_1} \cap \cdots \cap A_{i_k}))\right)^{(-1)^{k-1}} - 1\right). \tag{2.15}$$

Proof By utilizing the definition of $Z_{n,\lambda}^{(k)}$, Eq. (2.15) can be written as

$$Q_\lambda\left(\bigcup_{i=1}^{n} A_i\right) = \frac{1}{\lambda}\left(\prod_{k=1}^{n}\left(Z_{n,\lambda}^{(k)}\right)^{(-1)^{k-1}} - 1\right). \tag{2.16}$$

By making the use of Proposition 2.4, it can be shown by direct calculation that the formula in Eq. (2.16) holds for $n = 2, 3$; that is,

$$Q_\lambda(A_1 \cup A_2) = \frac{1}{\lambda}\left(\left(Z_{2,\lambda}^{(1)}\right)\left(Z_{2,\lambda}^{(2)}\right)^{-1} - 1\right)$$

$$Q_\lambda(A_1 \cup A_2 \cup A_3) = \frac{1}{\lambda}\left(\left(Z_{3,\lambda}^{(1)}\right)\left(Z_{3,\lambda}^{(2)}\right)^{-1}\left(Z_{3,\lambda}^{(3)}\right) - 1\right).$$

Here, we will apply induction; that is, we will prove that if Eq. (2.16) holds, then the equation

$$Q_\lambda\left(\bigcup_{i=1}^{n+1} A_i\right) = \frac{1}{\lambda}\left(\prod_{k=1}^{n+1}\left(Z_{n+1,\lambda}^{(k)}\right)^{(-1)^{k-1}} - 1\right) \tag{2.17}$$

holds as well.

By making use of Lemma 2.4, the associativity of the set union operation and the distributivity of the set intersection operation over the set union operation, we have

$$\begin{aligned}
Q_\lambda\left(\bigcup_{i=1}^{n+1} A_i\right) &= Q_\lambda\left(\left(\bigcup_{i=1}^{n} A_i\right) \cup A_{n+1}\right) = \\
&= \frac{1}{1+\lambda Q_\lambda\left(\left(\bigcup_{i=1}^{n} A_i\right) \cap A_{n+1}\right)}\left(Q_\lambda\left(\bigcup_{i=1}^{n} A_i\right) + Q_\lambda(A_{n+1}) + \right. \\
&\left. + \lambda Q_\lambda\left(\bigcup_{i=1}^{n} A_i\right) Q_\lambda(A_{n+1}) - Q_\lambda\left(\left(\bigcup_{i=1}^{n} A_i\right) \cap A_{n+1}\right)\right) = \\
&= \frac{1}{1+\lambda Q_\lambda\left(\bigcup_{i=1}^{n} (A_i \cap A_{n+1})\right)}\left(Q_\lambda\left(\bigcup_{i=1}^{n} A_i\right) + Q_\lambda(A_{n+1}) + \right. \\
&\left. + \lambda Q_\lambda\left(\bigcup_{i=1}^{n} A_i\right) Q_\lambda(A_{n+1}) - Q_\lambda\left(\bigcup_{i=1}^{n} (A_i \cap A_{n+1})\right)\right).
\end{aligned} \tag{2.18}$$

Now, by introducing $A_i^* = A_i \cap A_{n+1}$ for all $1 \le i \le n$, Eq. (2.18) can be written as

$$Q_\lambda\left(\bigcup_{i=1}^{n+1} A_i\right) = \frac{1}{1+\lambda Q_\lambda\left(\bigcup_{i=1}^{n} A_i^*\right)}\left(Q_\lambda\left(\bigcup_{i=1}^{n} A_i\right) + Q_\lambda(A_{n+1}) + \right.$$
$$\left. + \lambda Q_\lambda\left(\bigcup_{i=1}^{n} A_i\right) Q_\lambda(A_{n+1}) - Q_\lambda\left(\bigcup_{i=1}^{n} A_i^*\right)\right). \tag{2.19}$$

Next, using the inductive condition and the fact that $Z_{n,\lambda}^{(k)} = p_{n,\lambda}^{(k)}(A_1, \ldots, A_n)$ holds by definition for any $1 \leq k \leq n$, $Q_\lambda\left(\bigcup_{i=1}^{n} A_i^*\right)$ can be written as

$$Q_\lambda\left(\bigcup_{i=1}^{n} A_i^*\right) = \frac{1}{\lambda}\left(\prod_{k=1}^{n}\left(p_{n,\lambda}^{(k)}\left(A_1^*, \ldots, A_n^*\right)\right)^{(-1)^{k-1}} - 1\right).$$

Since $Z_{n,\lambda}^{*(k)} = p_{n,\lambda}^{(k)}(A_1^*, \ldots, A_n^*)$ holds by definition for any $1 \leq k \leq n$, the previous equation can be rewritten in the following form:

$$Q_\lambda\left(\bigcup_{i=1}^{n} A_i^*\right) = \frac{1}{\lambda}\left(\prod_{k=1}^{n}\left(Z_{n,\lambda}^{*(k)}\right)^{(-1)^{k-1}} - 1\right). \tag{2.20}$$

Recall that based on Lemma 2.3, we have the following equation

$$Z_{n,\lambda}^{*(k)} = \begin{cases} \frac{Z_{n+1,\lambda}^{(k+1)}}{Z_{n,\lambda}^{(k+1)}}, & \text{if } k < n \\ Z_{n+1,\lambda}^{(n+1)}, & \text{if } k = n. \end{cases} \tag{2.21}$$

Applying Eq. (2.21) to Eq. (2.20) yields

$$Q_\lambda\left(\bigcup_{i=1}^{n} A_i^*\right) =$$
$$= \frac{1}{\lambda}\left(\left(\prod_{k=1}^{n-1}\left(Z_{n+1,\lambda}^{(k+1)}\right)^{(-1)^{k-1}}\left(Z_{n,\lambda}^{(k+1)}\right)^{(-1)^{k}}\right)\left(Z_{n+1,\lambda}^{(n+1)}\right)^{(-1)^{n+1}} - 1\right). \tag{2.22}$$

Next, based on Lemma 2.4,

$$Q_\lambda(A_{n+1}) = \frac{1}{\lambda}\left(\frac{Z_{n+1,\lambda}^{(1)}}{Z_{n,\lambda}^{(1)}} - 1\right). \tag{2.23}$$

Now, applying the inductive condition given in Eq. (2.16) and substituting the formulas for $Q_\lambda\left(\bigcup_{i=1}^{n} A_i^*\right)$ and $Q_\lambda(A_n)$ given by Eqs. (2.22) and (2.23), respectively, into Eq. (2.19) gives us

$$Q_\lambda\left(\bigcup_{i=1}^{n+1} A_i\right) = \\ = \frac{\frac{1}{\lambda}(Y_1-1)+\frac{1}{\lambda}(Y_2-1)+\lambda\frac{1}{\lambda}(Y_1-1)\frac{1}{\lambda}(Y_2-1)-\frac{1}{\lambda}(Y_3-1)}{Y_3}, \tag{2.24}$$

where

$$Y_1 = \prod_{k=1}^{n}\left(Z_{n,\lambda}^{(k)}\right)^{(-1)^{k-1}}$$

$$Y_2 = \frac{Z_{n+1,\lambda}^{(1)}}{Z_{n,\lambda}^{(1)}}$$

$$Y_3 = \left(\prod_{k=1}^{n-1}\left(Z_{n+1,\lambda}^{(k+1)}\right)^{(-1)^{k-1}}\left(Z_{n,\lambda}^{(k+1)}\right)^{(-1)^{k}}\right)\left(Z_{n+1,\lambda}^{(n+1)}\right)^{(-1)^{n+1}}.$$

Simplifying Eq. (2.24) leads to

$$Q_\lambda\left(\bigcup_{i=1}^{n+1} A_i\right) = \frac{1}{\lambda}\left(\frac{Y_1 Y_2}{Y_3} - 1\right). \tag{2.25}$$

Now, by substituting the definitions of Y_1, Y_2 and Y_3 into Eq. (2.25), after simplifications we get

$$Q_\lambda\left(\bigcup_{i=1}^{n+1} A_i\right) = \\ = \frac{1}{\lambda}\left(\frac{\left(\prod_{k=1}^{n}\left(Z_{n,\lambda}^{(k)}\right)^{(-1)^{k-1}}\right)\frac{Z_{n+1,\lambda}^{(1)}}{Z_{n,\lambda}^{(1)}}}{\left(\prod_{k=1}^{n-1}\left(Z_{n+1,\lambda}^{(k+1)}\right)^{(-1)^{k-1}}\right)\left(\prod_{k=1}^{n-1}\left(Z_{n,\lambda}^{(k+1)}\right)^{(-1)^{k}}\right)\left(Z_{n+1,\lambda}^{(n+1)}\right)^{(-1)^{n+1}}} - 1\right) = \\ = \frac{1}{\lambda}\left(\frac{Z_{n+1,\lambda}^{(1)}}{\left(\prod_{k=1}^{n-1}\left(Z_{n+1,\lambda}^{(k+1)}\right)^{(-1)^{k-1}}\right)\left(Z_{n+1,\lambda}^{(n+1)}\right)^{(-1)^{n+1}}} - 1\right) = \\ = \frac{1}{\lambda}\left(\frac{Z_{n+1,\lambda}^{(1)}}{\left(Z_{n+1,\lambda}^{(2)}\right)\left(Z_{n+1,\lambda}^{(3)}\right)^{-1}\cdots\left(Z_{n+1,\lambda}^{(n+1)}\right)^{(-1)^{n+1}}} - 1\right) = \\ = \frac{1}{\lambda}\left(\left(Z_{n+1,\lambda}^{(1)}\right)\left(Z_{n+1,\lambda}^{(2)}\right)^{-1}\left(Z_{n+1,\lambda}^{(3)}\right)\cdots\left(Z_{n+1,\lambda}^{(n+1)}\right)^{(-1)^{n}} - 1\right) = \\ = \frac{1}{\lambda}\left(\prod_{k=1}^{n+1}\left(Z_{n+1,\lambda}^{(k)}\right)^{(-1)^{k-1}} - 1\right).$$

Notice that this formula is the same as the formula for $Q_\lambda\left(\bigcup_{i=1}^{n+1} A_i\right)$ given in Eq. (2.17), which means that we have proved this theorem. □

On the one hand, Theorem 2.1 tells us how to compute the λ-additive measure of the union of n sets in the case where λ is nonzero. On the other hand, it immediately follows from the definition of λ-additive measure that if $\lambda = 0$, then the λ-additive measure on the finite set X is a probability measure on X. Hence, if X is a finite set, $A_1, \ldots, A_n \in \mathcal{P}(X)$, $n \geq 2$, Q_λ is a λ-additive measure on X and $\lambda = 0$, then $Q_\lambda\left(\bigcup_{i=1}^{n} A_i\right)$ can be computed by using the Poincaré formula of probability theory:

$$Q_\lambda\left(\bigcup_{i=1}^{n} A_i\right) = \sum_{k=1}^{n}(-1)^{k-1} \sum_{1 \leq i_1 < \cdots < i_k \leq n} Q_\lambda\left(A_{i_1} \cap \cdots \cap A_{i_k}\right). \tag{2.26}$$

The following theorem demonstrates that the Poincaré formula of probability theory given in Eq. (2.26) may be viewed as a special case of the general formula of λ-additive measure of union of n sets given in Eq. (2.15).

Theorem 2.2 *If X is a finite set, $A_1, \ldots, A_n \in \mathcal{P}(X)$, $n \geq 2$, Q_λ is a λ-additive measure on X and $\lambda \neq 0$, then*

$$\lim_{\lambda \to 0} Q_\lambda\left(\bigcup_{i=1}^{n} A_i\right) = \sum_{k=1}^{n}(-1)^{k-1} \sum_{1 \leq i_1 < \cdots < i_k \leq n} Q_\lambda\left(A_{i_1} \cap \cdots \cap A_{i_k}\right). \tag{2.27}$$

Proof Let $\lambda \neq 0$. Here, we will distinguish two cases: (1) when n is even, (2) when n is odd.

(1) If n is even, then based on Theorem 2.1,

$$\begin{aligned}\lim_{\lambda \to 0} Q_\lambda\left(\bigcup_{i=1}^{n} A_i\right) &= \lim_{\lambda \to 0}\left(\frac{1}{\lambda}\left(\frac{Z_{n,\lambda}^{(1)} Z_{n,\lambda}^{(3)} \cdots Z_{n,\lambda}^{(n-1)}}{Z_{n,\lambda}^{(2)} Z_{n,\lambda}^{(4)} \cdots Z_{n,\lambda}^{(n)}} - 1\right)\right) = \\ &= \frac{\lim_{\lambda \to 0}\left(\frac{1}{\lambda}\left(Z_{n,\lambda}^{(1)} Z_{n,\lambda}^{(3)} \cdots Z_{n,\lambda}^{(n-1)} - Z_{n,\lambda}^{(2)} Z_{n,\lambda}^{(4)} \cdots Z_{n,\lambda}^{(n)}\right)\right)}{\lim_{\lambda \to 0}\left(Z_{n,\lambda}^{(2)} Z_{n,\lambda}^{(4)} \cdots Z_{n,\lambda}^{(n)}\right)},\end{aligned} \tag{2.28}$$

where

$$Z_{n,\lambda}^{(k)} = \prod_{1 \leq i_1 < \cdots < i_k \leq n}\left(1 + \lambda Q_\lambda\left(A_{i_1} \cap \cdots \cap A_{i_k}\right)\right),$$

$1 \leq k \leq n$. Definition of $Z_{n,\lambda}^{(k)}$ implies that

$$\lim_{\lambda \to 0}\left(Z_{n,\lambda}^{(2)} Z_{n,\lambda}^{(4)} \cdots Z_{n,\lambda}^{(n)}\right) = 1. \tag{2.29}$$

Let $F(\lambda; A_1, \ldots, A_n) = Z^{(1)}_{n,\lambda} Z^{(3)}_{n,\lambda} \cdots Z^{(n-1)}_{n,\lambda} - Z^{(2)}_{n,\lambda} Z^{(4)}_{n,\lambda} \cdots Z^{(n)}_{n,\lambda}$. Applying the definition of $Z^{(k)}_{n,\lambda}$, after direct calculations we get

$$\begin{aligned} F(\lambda; A_1, \ldots, A_n) = \\ 1 + \sum_{1 \le i \le n} \lambda Q_\lambda(A_i) + \sum_{1 \le i_1 < i_2 < i_3 \le n} \lambda Q_\lambda(A_{i_1} \cap A_{i_2} \cap A_{i_3}) + \cdots \\ \cdots + \sum_{1 \le i_1 < \cdots < i_{n-1} \le n} \lambda Q_\lambda(A_{i_1} \cap \cdots \cap A_{i_{n-1}}) + G(\lambda) - \\ -1 - \sum_{1 \le i_1 < i_2 \le n} \lambda Q_\lambda(A_{i_1} \cap A_{i_2}) - \sum_{1 \le i_1 < i_2 < i_3 < i_4 \le n} \lambda Q_\lambda(A_{i_1} \cap A_{i_2} \cap A_{i_3} \cap A_{i_4}) - \cdots \\ \cdots - \sum_{1 \le i_1 < \cdots < i_n \le n} \lambda Q_\lambda(A_{i_1} \cap \cdots \cap A_{i_n}) - H(\lambda), \end{aligned}$$

where $G(\lambda)$ and $H(\lambda)$ are at least second order polynomials of λ in which the constant term is equal to zero. Thus,

$$\lim_{\lambda \to 0} \left(\frac{1}{\lambda} G(\lambda) \right) = 0, \qquad \lim_{\lambda \to 0} \left(\frac{1}{\lambda} H(\lambda) \right) = 0$$

and so

$$\begin{aligned} \lim_{\lambda \to 0} \left(\frac{1}{\lambda} F(\lambda; A_1, \ldots, A_n) \right) = \\ = \sum_{1 \le i \le n} Q_\lambda(A_i) - \sum_{1 \le i_1 < i_2 \le n} Q_\lambda(A_{i_1} \cap A_{i_2}) + \\ + \sum_{1 \le i_1 < i_2 < i_3 \le n} Q_\lambda(A_{i_1} \cap A_{i_2} \cap A_{i_3}) - \sum_{1 \le i_1 < i_2 < i_3 < i_4 \le n} Q_\lambda(A_{i_1} \cap A_{i_2} \cap A_{i_3} \cap A_{i_4}) + \\ \cdots \\ + \sum_{1 \le i_1 < \cdots < i_{n-1} \le n} Q_\lambda(A_{i_1} \cap \cdots \cap A_{i_{n-1}}) - \sum_{1 \le i_1 < \cdots < i_n \le n} Q_\lambda(A_{i_1} \cap \cdots \cap A_{i_n}). \end{aligned}$$

That is, we have the following equation:

$$\begin{aligned} \lim_{\lambda \to 0} \left(\frac{1}{\lambda} F(\lambda; A_1, \ldots, A_n) \right) = \\ = \lim_{\lambda \to 0} \left(\frac{1}{\lambda} \left(Z^{(1)}_{n,\lambda} Z^{(3)}_{n,\lambda} \cdots Z^{(n-1)}_{n,\lambda} - Z^{(2)}_{n,\lambda} Z^{(4)}_{n,\lambda} \cdots Z^{(n)}_{n,\lambda} \right) \right) = \\ = \sum_{k=1}^{n} (-1)^{k-1} \sum_{1 \le i_1 < \cdots < i_k \le n} Q_\lambda \left(A_{i_1} \cap \cdots \cap A_{i_k} \right). \end{aligned} \tag{2.30}$$

Now, by substituting the formulas in Eqs. (2.29) and (2.30) into Eq. (2.28), we get

$$\lim_{\lambda \to 0} Q_\lambda \left(\bigcup_{i=1}^{n} A_i \right) = \sum_{k=1}^{n} (-1)^{k-1} \sum_{1 \le i_1 < \cdots < i_k \le n} Q_\lambda \left(A_{i_1} \cap \cdots \cap A_{i_k} \right).$$

(2) In the case where n is an odd number, the theorem can be proven by following steps similar to those of case (1). □

This result tells us that our general formula for the λ-additive measure of the union of n sets may be viewed as the generalization of the Poincaré formula of probability theory.

Remark 2.2 It is worth mentioning here that the computation of the λ-additive measure of collection of pairwise disjoint sets can be derived from the general Poincaré formula in Eq. (2.15). Namely, since $A_1, A_2, \ldots, A_n \in \mathcal{P}(X)$ are pairwise disjoint sets, $Q_\lambda\left(A_{i_1} \cap \cdots \cap A_{i_k}\right) = 0$ holds for any $2 \leq k \leq n$; and so, if $\lambda > -1, \lambda \neq 0$, then by utilizing Theorem 2.1, we immediately get

$$Q_\lambda\left(\bigcup_{i=1}^{n} A_i\right) = \frac{1}{\lambda}\left(\prod_{i=1}^{n}(1 + \lambda Q_\lambda(A_i)) - 1\right). \tag{2.31}$$

We will use the following definition to demonstrate some basic properties of the general formula for λ-additive measure of the union of n sets.

Definition 2.4 Let X be a finite set, $A_1, \ldots, A_n \in \mathcal{P}(X)$, $n \geq 2$, Q_λ a λ-additive measure on X and let $\lambda \neq 0$. Then, the function $g : \mathcal{P}^n(X) \to \mathbb{R}$ is given by

$$g(A_1, \ldots, A_n) = \frac{1}{\lambda}\left(\prod_{k=1}^{n}\left(Z_{n,\lambda}^{(k)}\right)^{(-1)^{k-1}} - 1\right),$$

where

$$Z_{n,\lambda}^{(k)} = \prod_{1 \leq i_1 < \cdots < i_k \leq n} \left(1 + \lambda Q_\lambda\left(A_{i_1} \cap \cdots \cap A_{i_k}\right)\right).$$

Notice that $g(A_1, \ldots, A_n)$ is identical to the right hand side of the general formula for λ-additive measure of the union of n sets given in Eq. (2.15). Here, we expect this formula to satisfy some basic requirements. Namely,

(1) if $A_1 = \cdots = A_n = A$, then $Q_\lambda\left(\bigcup_{i=1}^{n} A_i\right) = Q_\lambda(A)$; that is, we expect that $g(A, \ldots, A) = Q_\lambda(A)$
(2) if $A_1 = \cdots = A_n = \emptyset$, then $Q_\lambda\left(\bigcup_{i=1}^{n} A_i\right) = 0$; that is, we expect that $g(\emptyset, \ldots, \emptyset) = 0$
(3) if $A_n = \emptyset$, then $Q_\lambda\left(\bigcup_{i=1}^{n} A_i\right) = Q_\lambda\left(\bigcup_{i=1}^{n-1} A_i\right)$; that is, we expect that $g(A_1, \ldots, A_{n-1}, \emptyset) = g(A_1, \ldots, A_{n-1})$.

The following lemmas will demonstrate that the function g, and hence the general formula for λ-additive measure of the union of n sets given in Eq. (2.15), also meet the above three requirements.

Lemma 2.5 *If $A_1 = \cdots = A_n = A$, then $g(A, \ldots, A) = Q_\lambda(A)$.*

Proof If $A_1 = \cdots = A_n = A$, then

$$Z_{n,\lambda}^{(k)} = \prod_{1 \le i_1 < \cdots < i_k \le n} \left(1 + \lambda Q_\lambda \left(A_{i_1} \cap \cdots \cap A_{i_k}\right)\right) = (1 + \lambda Q_\lambda(A))^{\binom{n}{k}},$$

for all $k = 1, \ldots, n$ and so

$$g(A_1, \ldots, A_n) = \frac{1}{\lambda}\left((1 + \lambda Q_\lambda(A))^{\sum_{k=1}^{n}(-1)^{k-1}\binom{n}{k}} - 1\right). \tag{2.32}$$

Next, Lemma 1.3 implies that

$$\sum_{k=1}^{n}(-1)^{k-1}\binom{n}{k} = 1,$$

and substituting this result into Eq. (2.32), we get $g(A, \ldots, A) = Q_\lambda(A)$. □

Lemma 2.6 *If* $A_1 = \cdots = A_n = \emptyset$, *then* $g(A_1, \ldots, A_n) = 0$.

Proof Noting that $Q_\lambda(\emptyset) = 0$, if $A_1 = \cdots = A_n = \emptyset$, then

$$Z_{n,\lambda}^{(k)} = \prod_{1 \le i_1 < \cdots < i_k \le n} \left(1 + \lambda Q_\lambda \left(A_{i_1} \cap \cdots \cap A_{i_k}\right)\right) = 1$$

for all $k = 1, \ldots, n$. Hence, based on the definition of $g(A_1, \ldots, A_n)$, $g(A_1, \ldots, A_n) = 0$ trivially follows.

Lemma 2.7 *If* $n > 2$ *and* $A_n = \emptyset$, *then* $g(A_1, \ldots, A_n) = g(A_1, \ldots, A_{n-1})$.

Proof It follows from the the definition of $Z_{n,\lambda}^{(k)}$ that if $A_n = \emptyset$, then

$$Z_{n,\lambda}^{(k)} = \begin{cases} Z_{n-1,\lambda}^{(k)}, & \text{if } k < n \\ 1, & \text{if } k = n, \end{cases}$$

where $n > 2$. By making use of this result, we have

$$g(A_1, \ldots, A_n) = \frac{1}{\lambda}\left(\prod_{k=1}^{n}\left(Z_{n,\lambda}^{(k)}\right)^{(-1)^{k-1}} - 1\right) =$$

$$= \frac{1}{\lambda}\left(\prod_{k=1}^{n-1}\left(Z_{n-1,\lambda}^{(k)}\right)^{(-1)^{k-1}} - 1\right) = g(A_1, \ldots, A_{n-1}).$$

□

2.2 Connections Between λ-Additive Measures and Probability Measures

The next lemma concerns a basic connection between the λ-additive measure and the probability measure.

Lemma 2.8 *Let Q_λ be a λ-additive measure on the finite set X. Then, Q_λ is a probability measure on the set X if and only if $\lambda = 0$.*

Proof If $\lambda = 0$, then from Definition 2.1 we get that $Q_\lambda(A \cup B) = Q_\lambda(A) + Q_\lambda(B)$ holds for any $A, B \in \mathcal{P}(X)$ and $A \cap B = \emptyset$. This conclusion and the fact that $Q_\lambda(X) = 1$ imply that Q_λ is a probability measure.

If Q_λ is a probability measure, then $Q_\lambda(A \cup B) = Q_\lambda(A) + Q_\lambda(B)$ holds for any $A, B \in \mathcal{P}(X)$ and $A \cap B = \emptyset$. At the same time, $Q_\lambda(A \cup B) = Q_\lambda(A) + Q_\lambda(B) + \lambda Q_\lambda(A) Q_\lambda(B)$ holds as well, which implies that $\lambda = 0$. □

Here, we will demonstrate that the λ-additive measures can be utilized for generating probability measures; and, conversely, λ-additive measures can be generated from probability measures.

Definition 2.5 Let Σ be a σ-algebra over the set X. Then the function $\mu \colon \Sigma \to [0, \infty)$ is a measure on the space (X, Σ) if and only if μ satisfies the following requirements:

(1) $\forall A \in \Sigma : \mu(A) \geq 0$
(2) $\mu(\emptyset) = 0$
(3) $\forall A_1, A_2, \ldots \in \Sigma$, if $A_i \cap A_j = \emptyset, \forall i \neq j$, then

$$\mu\left(\bigcup_{i=1}^{\infty} A_i\right) = \sum_{i=1}^{\infty} \mu(A_i).$$

Proposition 2.5 *If Σ is a σ-algebra over the set X, Q_λ is a λ-additive measure, which satisfies the continuity property of monotone measures, on the space (X, Σ), $\lambda > -1$, $\lambda \neq 0$, $c > 0$ and the function $\hat{Q}_{\lambda,c} \colon \Sigma \to [0, \infty)$ is given by*

$$\hat{Q}_{\lambda,c}(A) = c \ln(1 + \lambda Q_\lambda(A))$$

for any $A \in \Sigma$, then $\hat{Q}_{\lambda,c}$ is a measure on the space (X, Σ).

Proof $\hat{Q}_{\lambda,c}(A)$ is trivially non-negative for any $A \in \Sigma$ and if $A = \emptyset$, then $\hat{Q}_{\lambda,c}(A) = 0$. That is, $\hat{Q}_{\lambda,c}$ satisfies requirements (1) and (2) of Definition 2.5. Next, let $A_1, A_2, \ldots \in \Sigma$ be a countable collection of pairwise disjoint sets. Now, utilizing the definition of $\hat{Q}_{\lambda,c}$, the fact that Q_λ is a λ-additive measure on (X, Σ) and Eq. (2.31), we get

$$\hat{Q}_{\lambda,c}\left(\bigcup_{i=1}^{\infty} A_i\right) = c \ln\left(1 + \lambda\left(Q_\lambda\left(\bigcup_{i=1}^{\infty} A_i\right)\right)\right) =$$
$$= c \ln\left(1 + \lambda\frac{1}{\lambda}\left(\prod_{i=1}^{\infty}(1 + \lambda Q_\lambda(A_i)) - 1\right)\right) =$$
$$= \sum_{i=0}^{\infty} c \ln(1 + \lambda Q(A_i)) = \sum_{i=0}^{\infty} \hat{Q}_{\lambda,c}(A_i).$$

It means that the function $\hat{Q}_{\lambda,c}$ satisfies requirement (3) in Definition 2.5 as well. □

Theorem 2.3 *Let the function* $h_\lambda\colon [0, 1] \to [0, 1]$ *be given by*

$$h_\lambda(x) = \begin{cases} \dfrac{(1+\lambda)^x - 1}{\lambda}, & \text{if } \lambda \neq 0 \\ x, & \text{if } \lambda = 0, \end{cases} \tag{2.33}$$

where $\lambda \in (-1, \infty)$*. Furthermore, let* Σ *be a* σ*-algebra over the set* X *and let* Q_λ *and* Pr *be two continuous functions on the space* (X, Σ) *such that*

$$Q_\lambda = h_\lambda \circ Pr. \tag{2.34}$$

Then, Pr *is a probability measure on* (X, Σ) *if and only if* Q_λ *is a* λ*-additive measure on* (X, Σ)*.*

Proof Firstly, we will show that if Eq. (2.33), Eq. (2.34) hold and Q_λ is a λ-additive measure on (X, Σ), then Pr is a probability measure on (X, Σ). Notice that h_λ is a bijection and its inverse function $h_\lambda^{-1}\colon [0, 1] \to [0, 1]$ is

$$h_\lambda^{-1}(y) = \begin{cases} \dfrac{\ln(1+\lambda y)}{\ln(1+\lambda)}, & \text{if } \lambda \neq 0 \\ y, & \text{if } \lambda = 0. \end{cases} \tag{2.35}$$

We can see that, for a fixed $x \in [0, 1]$, the function $\lambda \longmapsto h_\lambda(x)$ is continuous. The continuity of $\lambda \longmapsto h_\lambda(x)$ at $\lambda = 0$ means that

$$\lim_{\lambda \to 0} \frac{(1+\lambda)^x - 1}{\lambda} = x. \tag{2.36}$$

Also, from Eq. (2.34), we have

$$Pr = h_\lambda^{-1} \circ Q_\lambda. \tag{2.37}$$

Let Q_λ be a λ-additive measure on (X, Σ). If $\lambda = 0$, then by noting Eqs. (2.35) and (2.37), we have $Pr(A) = Q_\lambda(A)$ for any $A \in \Sigma$. Since for $\lambda = 0$ any λ-additive measure is a probability measure (see Definitions 1.3 and 2.1), we get that Pr is a probability measure on (X, Σ).

Now, let $\lambda \in (-1, \infty)$ and $\lambda \neq 0$. Then, by noting Eqs. (2.35) and (2.37), we can write

$$Pr(A) = \frac{\ln(1 + \lambda Q_\lambda(A))}{\ln(1 + \lambda)}$$

for any $A \in \Sigma$. Since for any $A \in \Sigma$, $Pr(A) = \hat{Q}_{\lambda,c}(A)$ with $c = 1/\ln(1 + \lambda)$, based on Proposition 2.5, Pr is a measure. Moreover, as $Q_\lambda(X) = 1$, $Pr(X) = 1$ holds as well; the function Pr satisfies all the requirements of a probability measure given in Definition 1.3.

Secondly, we will show that if Eq. (2.33), Eq. (2.34) hold and Pr is a probability measure on (X, Σ), then Q_λ is a λ-additive measure on (X, Σ). Let Pr be a probability measure on (X, Σ). If $\lambda = 0$, then by noting Eqs. (2.33) and (2.34), we have $Q_\lambda(A) = Pr(A)$ for any $A \in \Sigma$. Therefore, by noting Definitions 1.3 and 2.1, we see that Q_λ is a λ-additive measure with $\lambda = 0$.

Now, let $\lambda \in (-1, \infty)$ and $\lambda \neq 0$. Then, from Eqs. (2.33) and (2.34) we have

$$Q_\lambda(A) = \frac{1}{\lambda}\left((1 + \lambda)^{Pr(A)} - 1\right) \tag{2.38}$$

for any $A \in \Sigma$. Since Pr is a probability measure on (X, Σ), $Pr(X) = 1$; and so from Eq. (2.38) we get $Q_\lambda(X) = 1$. That is, Q_λ satisfies requirement (1) of the λ-additive measures given by Definition 2.1. Now, let $A, B \in \Sigma$ such that $A \cap B = \emptyset$. Then, as Pr is a probability measure on (X, Σ), the equation

$$Pr(A \cup B) = Pr(A) + Pr(B) \tag{2.39}$$

holds. Utilizing Eqs. (2.38) and (2.39), $Q_\lambda(A \cup B)$ can be written as

$$\begin{aligned}
Q_\lambda(A \cup B) &= \frac{1}{\lambda}\left((1 + \lambda)^{Pr(A \cup B)} - 1\right) = \\
&= \frac{1}{\lambda}\left((1 + \lambda)^{Pr(A)+Pr(B)} - 1\right) = \\
= \frac{1}{\lambda}\left((1 + \lambda)^{Pr(A)} - 1\right) &+ \frac{1}{\lambda}\left((1 + \lambda)^{Pr(B)} - 1\right) + \\
+ \lambda\frac{1}{\lambda}\left((1 + \lambda)^{Pr(A)} - 1\right) &\frac{1}{\lambda}\left((1 + \lambda)^{Pr(B)} - 1\right) = \\
&= Q_\lambda(A) + Q_\lambda(B) + \lambda Q_\lambda(A) Q_\lambda(B).
\end{aligned}$$

It means that Q_λ satisfies requirement (2) of the λ-additive measures given in Definition 2.1 as well; that is, Q_λ meets all the requirements of a λ-additive measure. □

Remark 2.3 The measure Pr is independent of the base of the logarithm because for any $A \in \Sigma$

$$\frac{\log_a(1+\lambda Q_\lambda(A))}{\log_a(1+\lambda)} = \frac{\frac{\log_s(1+\lambda Q_\lambda(A))}{\log_s(a)}}{\frac{\log_s(1+\lambda)}{\log_s(a)}} = \frac{\log_s(1+\lambda Q_\lambda(A))}{\log_s(1+\lambda)},$$

where $a, s > 0$, $a, s \neq 1$, $\lambda > -1$ and $\lambda \neq 0$. Also, if $s = 1+\lambda$, then $Pr(A) = \log_{1+\lambda}(1+\lambda Q_\lambda(A))$.

2.2.1 The General Poincaré Formula Revisited

In Sect. 2.1.1.1, we introduced the general Poincaré formula for λ-additive measures and gave an elementary proof of it (see Theorem 2.1). Now, we will show how the results related to the connection between the λ-additive- and probability measures can be used to prove the Poincaré formula for λ-additive measures. In order to prove the next results, let us consider a fixed $\lambda \in (-1, \infty)$ and the corresponding strictly increasing bijection $h_\lambda \colon [0, 1] \to [0, 1]$, given via

$$h_\lambda(x) = \begin{cases} \dfrac{(1+\lambda)^x - 1}{\lambda}, & \text{if } \lambda \neq 0 \\ x, & \text{if } \lambda = 0 \end{cases}$$

with inverse $h_\lambda^{-1} \colon [0, 1] \to [0, 1]$, given via

$$h_\lambda^{-1}(y) = \begin{cases} \dfrac{\ln(1+\lambda y)}{\ln(1+\lambda)}, & \text{if } \lambda \neq 0 \\ y, & \text{if } \lambda = 0. \end{cases}$$

Now, let us consider some fixed $\lambda \in (-1, \infty)$, $\lambda \neq 0$ and a fixed λ-additive measure $Q_\lambda \colon \mathcal{P}(X) \to [0, 1]$. According to Theorem 2.3 (see also [15]), Q_λ is representable. More precisely, one has $Q_\lambda = h_\lambda \circ \mu$ for a uniquely determined additive measure $\mu \colon \mathcal{P}(X) \to [0, 1]$. With this in mind, we can prove the following theorem.

Theorem 2.4 *If X is a finite set, $A_1, \ldots, A_n \in \mathcal{P}(X)$, $n \geq 2$, Q_λ is a λ-additive measure on X and $\lambda \neq 0$, then*

$$Q_\lambda\left(\bigcup_{i=1}^{n} A_i\right) =$$

$$= \frac{1}{\lambda}\left(\prod_{k=1}^{n}\left(\prod_{1 \leq i_1 < \cdots < i_k \leq n} (1 + \lambda Q_\lambda(A_{i_1} \cap \cdots \cap A_{i_k}))\right)^{(-1)^{k-1}} - 1\right). \quad (2.40)$$

Proof Noting the Poincaré formula for probability measures, one has

$$\mu(A) = \sum_{k=1}^{n} (-1)^{k-1} a_k, \tag{2.41}$$

where $A \stackrel{def}{=} A_1 \cup A_2 \cup \cdots \cup A_n$ and $a_k \stackrel{def}{=} \sum_{1 \leq i_1 < \cdots < i_k \leq n} \mu\left(A_{i_1} \cap \cdots \cap A_{i_k}\right)$. Applying h_λ in both members of Eq. (2.41), we get

$$\begin{aligned} h_\lambda(\mu(A)) = Q_\lambda(A) = h_\lambda\left(\sum_{k=1}^{n} (-1)^{k-1} a_k\right) = \\ = \frac{(1+\lambda)^{\sum_{k=1}^{n}(-1)^{k-1} a_k} - 1}{\lambda} = \frac{\prod_{k=1}^{n} ((1+\lambda)^{a_k})^{(-1)^{k-1}} - 1}{\lambda}. \end{aligned} \tag{2.42}$$

It can be seen that

$$\begin{aligned} (1+\lambda)^{a_k} = \prod_{1 \leq i_1 < \cdots < i_k \leq n} (1+\lambda)^{\mu(A_{i_1} \cap \cdots \cap A_{i_k})} = \\ = \prod_{1 \leq i_1 < \cdots < i_k \leq n} \left(1 + \lambda Q_\lambda\left(A_{i_1} \cap \cdots \cap A_{i_k}\right)\right) \end{aligned}$$

because of the identity (valid for any $B \subset X$)

$$(1+\lambda)^{\mu(B)} = 1 + \lambda Q_\lambda(B).$$

Applying this to Eq. (2.42), we get Eq. (2.40). □

Remark 2.4 The Poincaré formula can be viewed as the limit case of the formula in Eq. (2.15) when λ tends to zero. Namely, in view of Eq. (2.36), one has for any $A = A_1 \cup A_2 \cup \cdots \cup A_n$:

$$\lim_{\lambda \to 0} Q_\lambda(A) = \lim_{\lambda \to 0} h_\lambda(\mu(A)) = \mu(A).$$

2.2.2 The Conditional λ-Additive Measure

Using the connection between the λ-additive measures and probability measures, Wenxiu and Lushu [18] introduced the conditional λ-additive measure. Here, following their approach and using our previous notations and constructions, we will give a brief description of the conditional λ-additive measure.

Definition 2.6 Let Σ be a σ-algebra on the set X and let $Q_\lambda\colon \Sigma \to [0, 1]$ be a λ-additive measure on the measurable space (X, Σ). For a fixed $\lambda > -1$, the conditional λ-additive measure $Q_\lambda(A|B)$; that is the conditional λ-additive measure of $A \in \Sigma$ given $B \in \Sigma$ and $Q_\lambda(B) \neq 0$ is defined as

$$Q_\lambda(A|B) = \begin{cases} \dfrac{(1+\lambda)^{\frac{\log_{1+\lambda}(1+\lambda Q_\lambda(A\cap B))}{\log_{1+\lambda}(1+\lambda Q_\lambda(B))}} - 1}{\lambda}, & \text{if } \lambda \neq 0 \\ \dfrac{Q_\lambda(A\cap B)}{Q_\lambda(B)}, & \text{if } \lambda = 0. \end{cases} \tag{2.43}$$

Now, we will show that the set function defined in Eq. (2.43) is in fact a λ-additive measure.

Proposition 2.6 *Let Σ be a σ-algebra on the set X and let $Q_\lambda\colon \Sigma \to [0, 1]$ be a λ-additive measure on the measurable space (X, Σ). For a fixed $\lambda > -1$, the set function $Q_\lambda(\cdot|B)\colon \Sigma \to [0, 1]$, which is given by Eq. (2.43) is a λ-additive measure on the measurable space (X, Σ).*

Proof Let the function $h_\lambda\colon [0, 1] \to [0, 1]$ be given by Eq. (2.33). Using Theorem 2.3, the set function $Pr\colon \Sigma \to [0, 1]$, $Pr = h_\lambda^{-1} \circ Q_\lambda$ is a probability measure on the measurable space (X, Σ), where h_λ^{-1} is the inverse function of h_λ. Noting Remark 2.3, function h_λ^{-1} can be written as

$$h_\lambda^{-1}(x) = \begin{cases} \log_{1+\lambda}(1+\lambda x), & \text{if } \lambda \neq 0 \\ x, & \text{if } \lambda = 0, \end{cases} \tag{2.44}$$

where $x \in [0, 1]$. Therefore, by using Definition 2.6, $Q_\lambda(A|B)$ can be written as

$$Q_\lambda(A|B) = h_\lambda\left(\frac{h_\lambda^{-1}(Q_\lambda(A\cap B))}{h_\lambda^{-1}(Q_\lambda(B))}\right). \tag{2.45}$$

Noting that $h_\lambda^{-1}(Q_\lambda(A\cap B)) = Pr(A\cap B)$ and $h_\lambda^{-1}(Q_\lambda(B)) = Pr(B)$, we have

$$\frac{h_\lambda^{-1}(Q_\lambda(A\cap B))}{h_\lambda^{-1}(Q_\lambda(B))} = \frac{Pr(A\cap B)}{Pr(B)} = Pr(A|B),$$

where $Pr(A|B)$ is the conditional probability of A given B ($Pr(B) > 0$). Hence, using Eq. (2.45) we have

$$Q_\lambda(\cdot|B) = h_\lambda\left(Pr(\cdot|B)\right),$$

and since $Pr(\cdot|B)$ is a probability measure on the space (X, Σ), by noting Theorem 2.3, we get that $Q_\lambda(\cdot|B)$ is a λ-additive measure on the space (X, Σ). □

Remark 2.5 Note that from Eq. (2.45), we have

$$h_\lambda^{-1}\left(Q_\lambda(A|B)\right) = \frac{h_\lambda^{-1}(Q_\lambda(A \cap B))}{h_\lambda^{-1}(Q_\lambda(B))}. \tag{2.46}$$

We know that if $\lambda = 0$, then the λ-additive measure is a probability measure and h_λ^{-1} is the identity function. In this case, Eq. (2.46) is the well-known definition of conditional probability.

Now, we will show how $Q_\lambda(B|A)$ can be expressed in terms of $Q_\lambda(A|B)$, $Q_\lambda(A)$ and $Q_\lambda(B)$. This result may be viewed as the Bayes' formula for λ-additive measures.

Proposition 2.7 *Let Σ be a σ-algebra on the set X and let $Q_\lambda\colon \Sigma \to [0, 1]$ be a λ-additive measure on the measurable space (X, Σ). For a fixed $\lambda > -1$ and $A, B \in \Sigma$, if $Q_\lambda(A), Q_\lambda(B) > 0$, then the conditional λ-additive measure $Q_\lambda(B|A)$ can be expressed in terms of the conditional λ-additive measure $Q_\lambda(A|B)$ and the λ-additive measures $Q_\lambda(A)$ and $Q_\lambda(B)$ as follows:*

$$Q_\lambda(B|A) = \begin{cases} \dfrac{(1+\lambda)^{\frac{\log_{1+\lambda}(1+\lambda Q_\lambda(A|B))\log_{1+\lambda}(1+\lambda Q_\lambda(B))}{\log_{1+\lambda}(1+\lambda Q_\lambda(A))}} - 1}{\lambda}, & \text{if } \lambda \neq 0 \\ \dfrac{Q_\lambda(A|B)Q_\lambda(B)}{Q_\lambda(A)}, & \text{if } \lambda = 0. \end{cases} \tag{2.47}$$

Proof Let the function $h_\lambda\colon [0, 1] \to [0, 1]$ be given by Eq. (2.33). Then the inverse function of h_λ; that is, h_λ^{-1} is given by Eq. (2.44). Now, using Definition 2.6 and Remark 2.5, we can write

$$h_\lambda^{-1}\left(Q_\lambda(B|A)\right) = \frac{h_\lambda^{-1}(Q_\lambda(A \cap B))}{h_\lambda^{-1}(Q_\lambda(A))} =$$

$$= \frac{h_\lambda^{-1}(Q_\lambda(A \cap B))}{h_\lambda^{-1}(Q_\lambda(B))} \frac{h_\lambda^{-1}(Q_\lambda(B))}{h_\lambda^{-1}(Q_\lambda(A))} = h_\lambda^{-1}(Q_\lambda(A|B)) \frac{h_\lambda^{-1}(Q_\lambda(B))}{h_\lambda^{-1}(Q_\lambda(A))}.$$

Next, by applying h_λ to both sides of the previous equation, we get

$$Q_\lambda(B|A) = h_\lambda\left(h_\lambda^{-1}(Q_\lambda(A|B)) \frac{h_\lambda^{-1}(Q_\lambda(B))}{h_\lambda^{-1}(Q_\lambda(A))}\right), \tag{2.48}$$

which, considering h_λ and h_λ^{-1}, is identical with Eq. (2.47). □

Remark 2.6 Note that from Eq. (2.48), we have

$$h_\lambda^{-1}\left(Q_\lambda(B|A)\right) = h_\lambda^{-1}(Q_\lambda(A|B)) \frac{h_\lambda^{-1}(Q_\lambda(B))}{h_\lambda^{-1}(Q_\lambda(A))}. \tag{2.49}$$

If $\lambda = 0$, then the λ-additive measure is a probability measure and h_λ^{-1} is the identity function. In this case, Eq. (2.49) is the well-known probabilistic Bayes formula.

The following proposition may be treated as the Bayes' theorem for λ-additive measures.

Proposition 2.8 *Let Σ be a σ-algebra on the set X and let $Q_\lambda\colon \Sigma \to [0, 1]$ be a λ-additive measure on the measurable space (X, Σ). For a fixed $\lambda > -1$ and $A_1, A_2, \ldots, A_n, B \in \Sigma$, if $Q_\lambda(A_i), Q_\lambda(B) > 0$, then the conditional λ-additive measure $Q_\lambda(A_i|B)$ can be expressed in terms of the conditional λ-additive measure $Q_\lambda(B|A_i)$ and the λ-additive measure $Q_\lambda(A_i)$ as follows:*

$$Q_\lambda(A_i|B) = \\ = \begin{cases} \dfrac{(1+\lambda)^{\frac{\log_{1+\lambda}(1+\lambda Q_\lambda(B|A_i))\log_{1+\lambda}(1+\lambda Q_\lambda(A_i))}{\sum_{k=1}^{n}\log_{1+\lambda}(1+\lambda Q_\lambda(B|A_k))\log_{1+\lambda}(1+\lambda Q_\lambda(A_k))}} - 1}{\lambda}, & \text{if } \lambda \neq 0 \\ \dfrac{Q_\lambda(B|A_i)Q_\lambda(A_i)}{\sum_{k=1}^{n} Q_\lambda(B|A_k)Q_\lambda(A_k)}, & \text{if } \lambda = 0, \end{cases} \tag{2.50}$$

where $i \in \{1, 2, \ldots, n\}$.

Proof Here again, let the function $h_\lambda\colon [0, 1] \to [0, 1]$ be given by Eq. (2.33). Using Theorem 2.3, the set function $Pr\colon \Sigma \to [0, 1]$, $Pr = h_\lambda^{-1} \circ Q_\lambda$ is a probability measure on the measurable space (X, Σ), where h_λ^{-1} is the inverse function of h_λ. Considering the functions h_λ and h_λ^{-1}, by using the law of total probability and the probabilistic Bayes formula, the right hand side of Eq. (2.50) can be written as

$$h_\lambda\left(\frac{h_\lambda^{-1}(Q_\lambda(B|A_i))h_\lambda^{-1}(Q_\lambda(A_i))}{\sum_{k=1}^{n} h_\lambda^{-1}(Q_\lambda(B|A_k))h_\lambda^{-1}(Q_\lambda(A_k))}\right) = \\ = h_\lambda\left(\frac{Pr(B|A_i)Pr(A_i)}{\sum_{k=1}^{n} Pr(B|A_k)Pr(A_k)}\right) = \\ = h_\lambda\left(\frac{Pr(B|A_i)Pr(A_i)}{Pr(B)}\right) = h_\lambda\left(Pr(A_i|B)\right).$$

Noting Theorem 2.3, the last term in the previous equation is

$$h_\lambda\left(Pr(A_i|B)\right) = Q_\lambda(A_i|B).$$

Therefore,

$$Q_\lambda(A_i|B) = h_\lambda\left(\frac{h_\lambda^{-1}(Q_\lambda(B|A_i))h_\lambda^{-1}(Q_\lambda(A_i))}{\sum_{k=1}^{n} h_\lambda^{-1}(Q_\lambda(B|A_k))h_\lambda^{-1}(Q_\lambda(A_k))}\right), \tag{2.51}$$

which means that Eq. (2.50) holds. □

Remark 2.7 Note that from Eq. (2.51), we have

$$h_\lambda^{-1}\left(Q_\lambda(A_i|B)\right) = \frac{h_\lambda^{-1}(Q_\lambda(B|A_i))h_\lambda^{-1}(Q_\lambda(A_i))}{\sum_{k=1}^{n} h_\lambda^{-1}(Q_\lambda(B|A_k))h_\lambda^{-1}(Q_\lambda(A_k))}. \tag{2.52}$$

Since for $\lambda = 0$, the λ-additive measure is a probability measure and h_λ^{-1} is the identity function, Eq. (2.52) describes the well-known probabilistic Bayes theorem if $\lambda = 0$.

Using the concept of the conditional λ-additive measure, we can define the λ-additive independence. This approach is similar to that as the probabilistic independence is defined using the notion of conditional probability.

Definition 2.7 Let Σ be a σ-algebra on the set X and let $Q_\lambda \colon \Sigma \to [0, 1]$ be a λ-additive measure on the measurable space (X, Σ). For a fixed $\lambda > -1$, the sets $A, B \in \Sigma$ said to be Q_λ-independent if and only if

for $\lambda \neq 0$,

$$\log_{1+\lambda}\left(1 + \lambda Q_\lambda(A \cap B)\right) = \log_{1+\lambda}\left(1 + \lambda Q_\lambda(A)\right) \log_{1+\lambda}\left(1 + \lambda Q_\lambda(B)\right), \tag{2.53}$$

and for $\lambda = 0$,

$$Q_\lambda(A \cap B) = Q_\lambda(A) Q_\lambda(B). \tag{2.54}$$

Remark 2.8 Using the functions h_λ and h_λ^{-1} given by Eqs. (2.33) and (2.44), respectively, Eqs. (2.53) and (2.54) can be written into the following common form:

$$h_\lambda^{-1}\left(Q_\lambda(A \cap B)\right) = h_\lambda^{-1}(Q_\lambda(A))h_\lambda^{-1}(Q_\lambda(B)). \tag{2.55}$$

If $\lambda = 0$, then the λ-additive measure is a probability measure and h_λ^{-1} is the identity function. In this case, Eq. (2.55) is the well-known definition of the probabilistic independence.

The following proposition is concerning a connection between the λ-additive independence and the conditional λ-additive measure.

Proposition 2.9 *Let Σ be a σ-algebra on the set X and let $Q_\lambda \colon \Sigma \to [0, 1]$ be a λ-additive measure on the measurable space (X, Σ). For a fixed $\lambda > -1$, if $A, B \in \Sigma$ are Q_λ-independent and $Q_\lambda(B) > 0$, then*

$$Q_\lambda(A|B) = Q_\lambda(A). \tag{2.56}$$

Proof Let the functions h_λ and h_λ^{-1} be given by Eqs. (2.33) and (2.44), respectively. Then, the Q_λ independence of A and B means that

$$h_\lambda^{-1}\left(Q_\lambda(A \cap B)\right) = h_\lambda^{-1}(Q_\lambda(A))h_\lambda^{-1}(Q_\lambda(B)).$$

Noting that $Q_\lambda(B) > 0$, from the last equation we have

$$\frac{h_\lambda^{-1}(Q_\lambda(A \cap B))}{h_\lambda^{-1}(Q_\lambda(B))} = h_\lambda^{-1}(Q_\lambda(A)).$$

Applying h to both sides of this equation, we get

$$h_\lambda\left(\frac{h_\lambda^{-1}(Q_\lambda(A \cap B))}{h_\lambda^{-1}(Q_\lambda(B))}\right) = Q_\lambda(A). \tag{2.57}$$

Now, noting the definition of $Q_\lambda(A|B)$, Eq. (2.57) can be written as

$$Q_\lambda(A|B) = Q_\lambda(A).$$

□

2.2.3 λ-Additive Measure on the Real Line

Here, we will show how a λ-additive measure can be constructed on the real line.

Proposition 2.10 *Let* $\mathbb{S} = \{[a, b) : -\infty < a \le b < \infty\}$ *and let* $F : \mathbb{R} \to [0, 1]$ *be a probability distribution function; that is, F is left continuous,* $\lim_{x\to-\infty} F(x) = 0$ *and* $\lim_{x\to\infty} F(x) = 1$. *Then, the set function* $Q_\lambda : \mathbb{S} \to [0, 1]$,

$$Q_\lambda([a, b)) = \frac{F(b) - F(a)}{1 + \lambda F(a)}, \quad [a, b) \in \mathbb{S}, \lambda > -1 \tag{2.58}$$

is a λ-additive measure on $\mathbb{S}$.

Proof By exploiting the fact that F is a probability distribution function, from Eq. (2.58), we have

$$Q_\lambda(X) = Q_\lambda((-\infty, \infty)) = \lim_{\substack{a\to-\infty \\ b\to+\infty}} Q_\lambda([a, b)) = 1. \tag{2.59}$$

Next, for any $[a, b) \in \mathbb{S}$ and $[b, c) \in \mathbb{S}$, we have $[a, b) \cup [b, c) = [a, c) \in \mathbb{S}$. After direct calculation, we get

$$\begin{aligned}
Q_\lambda([a, b)) + Q_\lambda([b, c)) + \lambda Q_\lambda([a, b))Q_\lambda([b, c)) = \\
= Q_\lambda([a, b)) + Q_\lambda([b, c))\,(1 + \lambda Q_\lambda([a, b))) = \\
= \frac{F(b) - F(a)}{1 + \lambda F(a)} + \frac{F(c) - F(b)}{1 + \lambda F(b)}\left(1 + \lambda\frac{F(b) - F(a)}{1 + \lambda F(a)}\right) = \\
= \frac{F(c) - F(a)}{1 + \lambda F(a)} = Q_\lambda([a, c)).
\end{aligned} \tag{2.60}$$

Here, Eqs. (2.59) and (2.60) mean that the set function Q_λ meets the criteria for a λ-additive measure given in Definition 2.1. □

Proposition 2.10 tells us that using a probability distribution on the real line, we can generate a λ-additive measure. Note that in Eq. (2.58), the probability distribution F is called the distribution function of the λ-additive measure Q_λ.

The following proposition concerns another connection between the probability distribution functions and the λ-additive measures on the real line.

Proposition 2.11 *Let* $\mathbb{S} = \{[a, b) \colon -\infty < a \leq b < \infty\}$ *and for a fixed* $\lambda > -1$ *let* $Q_\lambda \colon \mathbb{S} \to [0, 1]$ *be a λ-additive measure on* $\mathbb{S}$. *Then, the function* $F \colon \mathbb{R} \to [0, 1]$,

$$F(x) = Q_\lambda((-\infty, x))$$

is a probability distribution function and for any $[a, b) \in \mathbb{S}$

$$Q_\lambda([a, b)) = \frac{F(b) - F(a)}{1 + \lambda F(a)}.$$

Proof If Q_λ is a λ-additive measure on $\mathbb{S}$, then using the properties of the λ-additive measure, we have

$$\lim_{x \to -\infty} F(x) = \lim_{x \to -\infty} Q_\lambda((-\infty, x)) = Q_\lambda(\emptyset) = 0$$
$$\lim_{x \to +\infty} F(x) = \lim_{x \to +\infty} Q_\lambda((-\infty, x)) = 1$$

and $F(x) = Q_\lambda((-\infty, x))$ is obviously left continuous. It means that F is a probability distribution function. Next, using Proposition 2.2, for any $[a, b) \in \mathbb{S}$, we have

$$Q_\lambda([a, b)) = Q_\lambda((-\infty, b) \setminus (-\infty, a)) = \\ = \frac{Q_\lambda((-\infty, b)) - Q_\lambda((-\infty, a))}{1 + \lambda Q_\lambda((-\infty, a))} = \frac{F(b) - F(a)}{1 + \lambda F(a)}. \tag{2.61}$$

□

Remark 2.9 We know that if $\lambda = 0$, then the λ-additive measure Q_λ is a probability measure. In line with this, if $\lambda = 0$, then Eq. (2.61) becomes $Q_\lambda([a, b)) = F(b) - F(a)$.

Note that the results of Propositions 2.10 and 2.11 can be generalized to Borel fields, see Theorem 5.3 in [3].

2.3 Dual λ-Additive Measures and Their Properties

Here we will introduce the dual measure of a λ-additive measure and show that it is a λ-additive measure as well.

Proposition 2.12 *Let Q_λ be a λ-additive measures on the finite set X. Then its dual measure $\mu: \mathcal{P}(X) \to [0, 1]$, which is given by*

$$\mu(A) = 1 - Q_\lambda(\overline{A})$$

for any $A \in \mathcal{P}(X)$, is a λ-additive measure with the parameter

$$\lambda' = -\frac{\lambda}{1+\lambda}$$

(see Corollary 4.5 in [3]).

Proof Let Q_λ be a λ-additive measure on X and let $A, B \in \mathcal{P}(X)$ such that $A \cap B = \emptyset$. Then, $X = \overline{A \cap B} = \overline{A} \cup \overline{B}$. Now, noting that $\mu(A) = 1 - Q_\lambda(\overline{A})$, the formula for the λ-additive measure of the intersection of two sets given by Eq. (2.7) and the fact that $Q_\lambda(\overline{A} \cup \overline{B}) = Q_\lambda(X) = 1$, we get

$$\begin{aligned}\mu(A \cup B) &= 1 - Q_\lambda(\overline{A \cup B}) = 1 - Q_\lambda(\overline{A} \cap \overline{B}) = \\ &= 1 - \frac{Q_\lambda(\overline{A}) + Q_\lambda(\overline{B}) + \lambda Q_\lambda(\overline{A}) Q_\lambda(\overline{B}) - Q_\lambda(\overline{A} \cup \overline{B})}{1 + \lambda Q_\lambda(\overline{A} \cup \overline{B})} = \\ &= 1 - \frac{1 - \mu(A) + 1 - \mu(B) + \lambda(1 - \mu(A))(1 - \mu(B)) - 1}{1 + \lambda} = \\ &= \mu(A) + \mu(B) - \frac{\lambda}{1+\lambda}\mu(A)\mu(B).\end{aligned}$$

That is, the equation

$$\mu(A \cup B) = \mu(A) + \mu(B) + \lambda' \mu(A)\mu(B)$$

holds for any $A, B \in \mathcal{P}(X)$, $A \cap B = \emptyset$. Next, by taking into account the fact that $\mu(\emptyset) = 1 - Q_\lambda(X) = 0$, we see that the measure μ meets the criteria for a λ-additive measure (with parameter λ') on X (see Definition 2.1). □

Following the result of Proposition 2.12, we will utilize the concept of the dual pair of λ-additive measures.

Definition 2.8 Let Q_{λ_1} and Q_{λ_2} be two λ-additive measures on the finite set X. Then, Q_{λ_1} and Q_{λ_2} are said to be a dual pair of λ-additive measures if and only if

$$Q_{\lambda_1}(A) + Q_{\lambda_2}(\overline{A}) = 1$$

holds for any $A \in \mathcal{P}(X)$, where $\lambda_1, \lambda_2 \in (-1, \infty)$.

Here, we will demonstrate some key properties of the λ-additive measure related to a dual pair of λ-additive measures.

Theorem 2.5 *Let Q_{λ_1} and Q_{λ_2} be two λ-additive measures on the finite set X. Then Q_{λ_1} and Q_{λ_2} are a dual pair of λ-additive measures if and only if*

$$\lambda_2 = -\frac{\lambda_1}{1 + \lambda_1},$$

where $\lambda_1, \lambda_2 \in (-1, \infty)$.

Proof If Q_{λ_1} and Q_{λ_2} are a dual pair of λ-additive measures on the finite set X, then, by noting Proposition 2.12, we immediately have $\lambda_2 = -\frac{\lambda_1}{1+\lambda_1}$.

Now, we will show that if $\lambda_2 = -\frac{\lambda_1}{1+\lambda_1}$, then Q_{λ_1} and Q_{λ_2} are a dual pair of λ-additive measures on X. Since Q_{λ_1} and Q_{λ_2} are two λ-additive measures, according to Theorem 2.3, we have

$$Q_{\lambda_1}(A) = \begin{cases} \frac{1}{\lambda_1}\left((1 + \lambda_1)^{Pr(A)} - 1\right), & \text{if } \lambda_1 \neq 0 \\ Pr(A), & \text{if } \lambda_1 = 0 \end{cases} \tag{2.62}$$

and

$$Q_{\lambda_2}(A) = \begin{cases} \frac{1}{\lambda_2}\left((1 + \lambda_2)^{Pr(A)} - 1\right), & \text{if } \lambda_2 \neq 0 \\ Pr(A), & \text{if } \lambda_2 = 0 \end{cases} \tag{2.63}$$

for any $A \in \mathcal{P}(X)$, where Pr is a probability measure on X and $\lambda_1, \lambda_2 \in (-1, \infty)$. Next, since Pr is a probability measure, we have

$$1 - Pr(\overline{A}) = Pr(A) \tag{2.64}$$

for any $A \in \mathcal{P}(X)$. Now, let

$$\lambda_2 = -\frac{\lambda_1}{1 + \lambda_1}.$$

Then, $\lambda_1 = 0$ implies $\lambda_2 = 0$ and, conversely, $\lambda_2 = 0$ implies $\lambda_1 = 0$. In such a case, by taking into account Eqs. (2.62)–(2.64), we immediately have

$$Q_{\lambda_1}(A) + Q_{\lambda_2}(\overline{A}) = Pr(A) + Pr(\overline{A}) = 1.$$

Now, let $\lambda_1, \lambda_2 \neq 0$. Then, by using Eqs. (2.62) and (2.63), we can write

$$Q_{\lambda_1}(A) + Q_{\lambda_2}(\overline{A}) =$$
$$= \frac{1}{\lambda_1}\left((1+\lambda_1)^{Pr(A)} - 1\right) + \frac{1}{\lambda_2}\left((1+\lambda_2)^{Pr(\overline{A})} - 1\right) =$$
$$= \frac{1}{\lambda_1}\left((1+\lambda_1)^{Pr(A)} - 1\right) + \frac{1}{-\frac{\lambda_1}{1+\lambda_1}}\left(\left(1 - \frac{\lambda_1}{1+\lambda_1}\right)^{Pr(\overline{A})} - 1\right) = \tag{2.65}$$
$$= \frac{1}{\lambda_1}\left((1+\lambda_1)^{Pr(A)} - 1\right) - \frac{1}{\lambda_1}\left((1+\lambda_1)^{1-Pr(\overline{A})} - (1+\lambda_1)\right).$$

Therefore, by noting Eq. (2.64), from Eq. (2.65), we get

$$Q_{\lambda_1}(A) + Q_{\lambda_2}(\overline{A}) =$$
$$= \frac{1}{\lambda_1}\left((1+\lambda_1)^{Pr(A)} - 1\right) - \frac{1}{\lambda_1}\left((1+\lambda_1)^{Pr(A)} - (1+\lambda_1)\right) =$$
$$= -\frac{1}{\lambda_1} + \frac{1+\lambda_1}{\lambda_1} = 1,$$

which means that Q_{λ_1} and Q_{λ_2} are a dual pair of λ-additive measures. □

Utilizing the definition of the dual pair of λ-additive measures, the following corollary can be stated.

Corollary 2.1 *Let Q_{λ_1} and Q_{λ_2} be a dual pair of λ-additive measures on the finite set X. Then, $\lambda_1 \in (-1, 0]$ if and only if $\lambda_2 \in [0, \infty)$.*

Proof Since $\lambda_2 = -\frac{\lambda_1}{1+\lambda_1}$ is a bijection from $(-1, 0]$ to $[0, \infty)$, this corollary follows from Theorem 2.5. □

2.3.1 Connection with Belief-, Probability- and Plausibility Measures

In Chap. 1, we discussed about the belief- and plausibility measures and pointed out that these measures can be effectively applied to modeling and describing uncertainty. Here, we will show how the λ-additive measures are connected with the belief-, probability- and plausibility measures.

Proposition 2.13 *Let Q_λ be a λ-additive measure on the finite set X. Then, Q_λ is a belief measure on X if and only if $\lambda \geq 0$ (see Theorem 4.21 in [3, 19]).*

Proof Firstly, we will show that if Q_λ is a λ-additive measure on the finite set X and $\lambda \geq 0$, then Q_λ is a belief measure on X. Let Q_λ be a λ-additive measure on the finite set $X = \{x_1, x_2, \ldots, x_n\}$ and let $\lambda \geq 0$. We will distinguish the following two cases: (1) $\lambda = 0$, (2) $\lambda > 0$. (1) if $\lambda = 0$, then, based on the definition of the λ-additive measure, Q_λ is a probability measure, and so Q_λ is a belief measure as well (see Remark 1.2). (2) When $\lambda > 0$, we define the set function $m : \mathcal{P}(X) \to [0, 1]$ as follows:

$$m(A) = \begin{cases} \lambda^{|A|-1} \prod_{x_i \in A} Q_\lambda(\{x_i\}) & \text{if } A \neq \emptyset \\ 0 & \text{if } A = \emptyset \end{cases}$$

for any $A \in \mathcal{P}(X)$. It immediately follows from definition of m that $m(A) \geq 0$ for any $A \in \mathcal{P}(X)$. Next, by noting Remark 2.2, we have

$$Q_\lambda(A) = Q_\lambda\left(\bigcup_{x_i \in A}\{x_i\}\right) = \frac{1}{\lambda}\left(\prod_{x_i \in A}(1 + \lambda Q_\lambda(\{x_i\})) - 1\right) =$$

$$= \frac{1}{\lambda} \sum_{B \subseteq A, B \neq \emptyset} \lambda^{|B|} \prod_{x_i \in B} Q_\lambda(\{x_i\}) = \sum_{B \subseteq A, B \neq \emptyset} \lambda^{|B|-1} \prod_{x_i \in B} Q_\lambda(\{x_i\}) = \tag{2.66}$$

$$= \sum_{B \subseteq A, B \neq \emptyset} m(B).$$

Since Q_λ is a λ-additive measure on X, $Q_\lambda(X) = 1$ and so we also have

$$\sum_{B \subseteq X} m(B) = 1.$$

From the last equation and from the fact that m is non-negative, we get that m is a basic probability assignment. Furthermore, noting the definition of the belief measure in Definition 1.8, Eq. (2.66) tells us that Q_λ is a belief measure induced from m.

Secondly, we will show that if Q_λ is a λ-additive measure on the finite set X and Q_λ is also a belief measure on X, then $\lambda \geq 0$. Let Q_λ be a λ-additive measure and also a belief measure on the finite set $X = \{x_1, x_2, \ldots, x_n\}$. We know that every probability measure on the finite set X is also a belief measure on the set X (see Remark 1.2). Thus, will distinguish the following two cases: (1) Q_λ is a belief- and probability measure on X, (2) Q_λ is a belief measure but not a probability measure on X. (1) Since in this case Q_λ is a probability measure and a λ-additive measure on the set X, it immediately follows from Lemma 2.8 that $\lambda = 0$. (2) As in this case Q_λ is not a probability measure, from Lemma 2.8 we get that $\lambda \neq 0$. Next, on the one hand, since Q_λ is a belief measure on X, there exists a basic probability assignment $m : \mathcal{P}(X) \to [0, 1]$ such that for any $A \in \mathcal{P}(X)$

$$Q_\lambda(A) = \sum_{B \subseteq A} m(B).$$

On the other hand, by noting that Q_λ is a λ-additive measure on X, we have

$$Q_\lambda(A) = \sum_{B \subseteq A, B \neq \emptyset} \lambda^{|B|-1} \prod_{x_i \in B} Q_\lambda(\{x_i\}),$$

and from the last two equations $\lambda \geq 0$ follows. □

The next theorem summarizes important connections between the λ-additive measure and the belief-, probability- and plausibility measures.

Theorem 2.6 *Let X be a finite set and let Q_λ be a λ-additive measure on X. Then, on set X, Q_λ is a*

(1) plausibility measure if and only if $-1 < \lambda \leq 0$
(2) probability measure if and only if $\lambda = 0$
(3) belief measure if and only if $\lambda \geq 0$

(see also [20]).

Proof Notice that we showed (2) in Lemma 2.8 and we demonstrated (3) in Proposition 2.13. Hence, we need to prove (1).

Firstly, we will show that if Q_λ is a λ-additive measure on the finite set X and $\lambda \leq 0$, then Q_λ is a plausibility measure on X. Now, let Q_λ be a λ-additive measure on the finite set X and let $\lambda \leq 0$. Furthermore, let λ' be given by

$$\lambda' = -\frac{\lambda}{1+\lambda} \tag{2.67}$$

and let $Q_{\lambda'}$ be a λ-additive measure on the set X with the parameter λ'. Then, based on Theorem 2.5, $Q_{\lambda'}$ is the dual λ-additive pair of Q_λ; that is, $Q_{\lambda'}(A) = 1 - Q_\lambda(\overline{A})$ for any $A \in \mathcal{P}(X)$. Noting the fact that $\lambda' \geq 0$, from Proposition 2.13, we have that $Q_{\lambda'}$ is a belief measure on X. Thus, by utilizing Proposition 1.3, we get that Q_λ is a plausibility measure on X.

Secondly, we will show that if Q_λ is a λ-additive measure on the finite set X and Q_λ is also a plausibility measure on X, then $\lambda \leq 0$. Now, let Q_λ be a λ-additive measure and also a plausibility measure on the finite set X. Again, let λ' be given by Eq. (2.67) and let $Q_{\lambda'}$ be a λ-additive measure on the set X with the parameter λ'. Then, based on Theorem 2.5, $Q_{\lambda'}$ is the dual λ-additive pair of Q_λ; that is, $Q_{\lambda'}(A) = 1 - Q_\lambda(\overline{A})$ for any $A \in \mathcal{P}(X)$. Now, by utilizing Proposition 1.3, we get that $Q_{\lambda'}$ is a belief measure on X, and by noting Proposition 2.13 we get that $\lambda' \geq 0$. Next, since $\lambda' \geq 0$, from Eq. (2.67) $\lambda \leq 0$ follows. □

Proposition 2.14 *Let Q_{λ_1} and Q_{λ_2} be two λ-additive measures on the finite set X. Then, Q_{λ_1} and Q_{λ_2} are a dual pair of belief- and plausibility measures on X if and only if they are a dual pair of λ-additive measures on X.*

Proof Firstly, we will show that if the condition of the proposition is satisfied and Q_{λ_1} and Q_{λ_2} are a dual pair of belief- and plausibility measures on X, then Q_{λ_1} and Q_{λ_2} are a dual pair of λ-additive measures on X. Let Q_{λ_1} and Q_{λ_2} be a dual pair of belief- and plausibility measures on X. Since, Q_{λ_1} and Q_{λ_2} are a dual pair; that is, $Q_{\lambda_2}(A) = 1 - Q_{\lambda_1}(\overline{A})$ holds for any $A \in \mathcal{P}(X)$, and Q_{λ_1} and Q_{λ_2} are λ-additive measures on X, they are also a dual pair of λ-additive measures on X.

Secondly, we will show that if Q_{λ_1} and Q_{λ_2} are a dual pair of λ-additive measures on X, then Q_{λ_1} and Q_{λ_2} are a dual pair of belief- and plausibility measures on X. Let Q_{λ_1} and Q_{λ_2} be a dual pair of λ-additive measures on X. Then, based on Corollary 2.1, either $\lambda_1 \in (-1, 0]$ and $\lambda_2 \in [0, \infty)$, or $\lambda_1 \in [0, \infty)$ and $\lambda_2 \in (-1, 0]$ holds. Now, utilizing Theorem 2.6, we get that either Q_{λ_1} is a plausibility measure and Q_{λ_2} is a belief measure, or Q_{λ_1} is a belief measure and Q_{λ_2} is a plausibility measure. Thus, noting that Q_{λ_1} and Q_{λ_2} are a dual pair of λ-additive measures on X, we may conclude that they are also a dual pair of belief- and plausibility measures on X. □

The following theorem is about the connections between the dual pairs of λ-additive measures and the dual pairs of belief- and plausibility measures.

Theorem 2.7 *Let Q_{λ_1} and Q_{λ_2} be two λ-additive measures on the finite set X. Then, Q_{λ_1} and Q_{λ_2} are a dual pair of belief- and plausibility measures on X if and only if*

$$\lambda_2 = -\frac{\lambda_1}{1 + \lambda_1},$$

where $\lambda_1, \lambda_2 \in (-1, \infty)$.

Proof Following Proposition 2.14, if Q_{λ_1} and Q_{λ_2} are two λ-additive measures on the finite set X, then Q_{λ_1} and Q_{λ_2} are a dual pair of belief- and plausibility measures on X if and only if they are a dual pair of λ-additive measures on X. Furthermore, based on Theorem 2.5, if Q_{λ_1} and Q_{λ_2} are two λ-additive measures on the finite set X, then Q_{λ_1} and Q_{λ_2} are a dual pair of λ-additive measures if and only if $\lambda_2 = -\frac{\lambda_1}{1+\lambda_1}$. Hence, this theorem follows from Proposition 2.14 and Theorem 2.5. □

2.3.2 An Application to Belief- and Plausibility Measures

Now, we will show how our formula for the λ-additive measure of union of n sets can be applied in the theory of belief- and plausibility measures. Previously, we pointed out that the Poincaré formula of probability theory may be viewed as a special case of our general formula for the λ-additive measure of the union of n sets. Now, based on Theorem 2.6, we may also conclude that our general formula for the λ-additive measure of the union of n sets given in Eq. (2.15) holds for belief- and plausibility measures as well. Namely,

(1) if Bl (Pl) is a belief measure (plausibility measure) represented by the λ-additive measure Q_λ with $\lambda > 0$ ($-1 < \lambda < 0$), then Eq. (2.15) may be viewed as the general formula for the Bl measure (Pl measure) of the union of n sets;
(2) if Bl (Pl) is a belief measure (plausibility measure) represented by the λ-additive measure Q_λ with $\lambda = 0$, then Bl (Pl) is a probability measure and so, based on Theorem 2.2, the Bl measure (Pl measure) of the union of n sets can be computed as the limit of the right hand side of Eq. (2.15), when $\lambda \to 0$.

Let X be a finite set, $X = \bigcup_{i=1}^{n} A_i$, $A_i \cap A_j = \emptyset$ for any $i \neq j$, and let the value of $Q_\lambda(A_i)$ be given, where $i, j = 1, \ldots, n$, Q_λ is a λ-additive measure on X and $A_1, \ldots, A_n \in \mathcal{P}(X)$. Furthermore, let us assume that the value of parameter λ is unknown. As $Q_\lambda\left(\bigcup_{i=1}^{n} A_i\right) = Q_\lambda(X) = 1$, based on Theorem 2.6, we know that

(1) if $\sum_{i=1}^{n} Q_\lambda(A_i) > 1$ then Q_λ is a plausibility measure; that is, $-1 < \lambda < 0$
(2) if $\sum_{i=1}^{n} Q_\lambda(A_i) = 1$ then Q_λ is a probability measure; that is, $\lambda = 0$
(3) if $\sum_{i=1}^{n} Q_\lambda(A_i) < 1$ then Q_λ is a belief measure; that is, $\lambda > 0$.

Notice that $\sum_{i=1}^{n} Q_\lambda(A_i) = 1$ immediately implies that $\lambda = 0$ and so Q_λ is a probability measure. At the same time, in the other two cases, from the value of $\sum_{i=1}^{n} Q_\lambda(A_i)$ we can infer only on the sign of λ. Now, we will demonstrate how the general formula for the λ-additive measure of the union of n sets can be utilized for characterizing the λ-additive measure in these two cases as well. That is, we will show that if $A_1, \ldots, A_n$ are pairwise disjoint sets such that $X = \bigcup_{i=1}^{n} A_i$ and the values $Q_\lambda(A_1), \ldots, Q_\lambda(A_n)$ are all known, then the value of parameter lambda can be unambiguously determined. Since $X = \bigcup_{i=1}^{n} A_i$, $Q_\lambda(X) = 1$ and $A_1, \ldots, A_n$ are pairwise disjoint sets, based on Remark 2.2, we have the equation

$$1 = Q_\lambda\left(\bigcup_{i=1}^{n} A_i\right) = \frac{1}{\lambda}\left(\prod_{i=1}^{n}\left(1 + \lambda Q_\lambda\left(A_i\right)\right) - 1\right), \tag{2.68}$$

in which $Q_\lambda(A_i)$ is known for all $i = 1, \ldots, n$, and the value of parameter λ is unknown. The following proposition demonstrates that Eq. (2.68) has only one root in the interval $(-1, 0) \cup (0, \infty)$.

Proposition 2.15 *If X is a finite set, Q_λ is a λ-additive measure on X, $\lambda > -1$, $\lambda \neq 0$, $A_1, A_2, \ldots, A_n \in \mathcal{P}(X)$ are pairwise disjoint sets such that $X = \bigcup_{i=1}^{n} A_i$, $Q_\lambda(A_i) < 1$, $Q_\lambda(A_i), \ldots, Q_\lambda(A_n)$ are all known and $i \in \{1, 2, \ldots, n\}$, then the equation*

$$\frac{1}{\lambda}\left(\prod_{i=1}^{n}(1 + \lambda Q_\lambda(A_i)) - 1\right) = 1 \tag{2.69}$$

has only one root for the parameter λ in the interval $(-1, 0) \cup (0, \infty)$ (see also Theorem 4.7 in [3]).

Proof This proof is based on the proof of a theorem connected with the multiplicative utility functions described by Keeney in [21, Appendix B]. Since $\lambda \neq 0$, Eq. (2.69) can be written as

$$\lambda + 1 = \prod_{i=1}^{n}(1 + \lambda z_i), \tag{2.70}$$

where $z_i = Q_\lambda(A_i)$, $i = 1, 2, \ldots, n$. Now, let $S = \sum_{i=1}^{n} z_i$ and let the polynomial $f(q)$ be given as

$$f(q) = q + 1 - \prod_{i=1}^{n}(1 + qz_i), \tag{2.71}$$

where $-1 \leq q < \infty$. From Eqs. (2.70) and (2.71), we get the following results:

$$f(\lambda) = 0, \; f(0) = 0, \; f(-1) = -\prod_{i=1}^{n}(1 - z_i) < 0.$$

The first derivative of function f is

$$f'(q) = \frac{\mathrm{d}f(q)}{\mathrm{d}q} = 1 - \sum_{i=1}^{n} z_i \prod_{i \neq j}(1 + z_j q),$$

from which we can see that $f'(q)$ is decreasing (with respect to q) in the interval $(-1, \infty)$,

$$f'(0) = 1 - \sum_{i=1}^{n} z_i = 1 - S \tag{2.72}$$

and

$$\lim_{q \to \infty} f'(q) = -\infty. \tag{2.73}$$

Here, we will distinguish three cases: (1) $S < 1$; (2) $S = 1$; (3) $S > 1$.

(1) Equation (2.72) implies that if $S < 1$, then $f'(0) > 0$. Since $f'(0) > 0$ and $f'(q)$ is decreasing in the interval $(-1, \infty)$, $f'(q)$ is positive in $(-1, 0)$. Therefore, $f'(q) = 0$ has no root in $(-1, 0)$. Based on Eq. (2.73), $f'(\infty) = -\infty$, and so $f'(q) = 0$ has a unique root q^* in $(0, \infty)$. Since $f(0) = 0$ and $f'(q) > 0$ in $(0, q^*)$, $f(q) = 0$ has no root in $(0, q^*)$. As $f(q^*) > 0$ and $f'(q)$ is negative and decreasing to $-\infty$ in (q^*, ∞), $f(q) = 0$ has a unique root q_0 in (q^*, ∞). Moreover, $f(q) > 0$ in $(0, q_0)$ and $f(q) < 0$ in (q_0, ∞); that is, the unique root q_0 is in $(0, \infty)$.
(2) It follows from Eq. (2.72) that if $S = 1$, then $f'(0) = 0$. Since $f'(0) = 0$ and $f'(q)$ is decreasing in the interval $(-1, \infty)$, $f'(q)$ is positive in the interval $(-1, 0)$ and it is negative in the interval $(0, \infty)$. Thus, $q = 0$ is the only root of $f'(q) = 0$ in the interval $(-1, \infty)$. Moreover, since $f(0) = 0, q = 0$ is the only root of $f(q) = 0$. It means that if $S = 1$, then the only solution of Eq. (2.70) is $\lambda = 0$. Recall that $\lambda \neq 0$; that is, in this case we do not get any solution to the equation in (2.69).
(3) Equation (2.72) implies that if $S > 1$, then $f'(0) < 0$. Since $f'(0) < 0$ and $f'(q)$ is decreasing in the interval $(-1, \infty)$, $f'(q)$ is negative in $(0, \infty)$. As $f(0) = 0$ and $f'(q)$ is negative in $(0, \infty)$, $f(q) = 0$ has no root in $(0, \infty)$. On the one hand, as $f(0) = 0$ and $f'(0) < 0$, $f(q) > 0$ immediately to the left of zero. On the other hand, $f(-1) < 0$. It means that there must be at least one root q_0 of

$f(q) = 0$ in $(-1, 0)$. Since $f'(q)$ is decreasing and $f(0) = 0$, q_0 is the unique root of $f(q) = 0$ in $(0, 1)$.

□

Based on this result, Eq. (2.68) can be numerically solved for λ. The value of parameter λ informs us about the 'plausibilitiness' or 'beliefness' of the Q_λ measure.

Example 2.2 Let $A_1, \ldots, A_n$ be n pairwise disjoint groups of people and let $X = \bigcup_{i=1}^n A_i$ be the universe of groups. Furthermore, let us assume that we have the value of $Q_\lambda(A_i)$ for all $i = 1, \ldots, n$, and the λ-additive measure is a performance measure; that is, $Q_\lambda(A_i)$ represents the performance of group A_i. Here, if $\sum_{i=1}^n Q_\lambda(A_i) = 1$, then Q_λ is a probability measure. If $\sum_{i=1}^n Q_\lambda(A_i) \neq 1$, then the solution of Eq. (2.68) for λ informs us about the 'plausibilitiness' or 'beliefness' of the measure Q_λ. Namely, if $\lambda \gg 0$, then Q_λ is a 'strong' belief measure, which indicates that uniting all the groups into one results in a better performing group. Similarly, if $-1 < \lambda \ll 0$, then Q_λ is a 'strong' plausibility measure, which tells us that merging all the groups into one results in a worse performing group.

2.4 Inequalities for λ-Additive Measures

2.4.1 *Characteristic Inequalities for Belief- and Plausibility Measures Using the λ-Additive Measure*

Now, we will present novel characteristic inequalities that are connected with the λ-additive-, belief- and plausibility measures. These inequalities, which are based on the application of the general Poincaré formula for λ-additive measures, may be viewed as new criteria for the belief- and plausibility measures when they are λ-additive measures as well.

Theorem 2.1, which states the general Poincaré formula for λ-additive measures, can also be given in the form of the following Proposition.

Proposition 2.16 *If X is a finite set, Q_λ is a λ-additive measure on X, $\lambda \in (-1, \infty)$, $\lambda \neq 0$, $A_1, \ldots, A_n \in \mathcal{P}(X)$ and $n \geq 2$, then*

$$Q_\lambda\left(\bigcup_{i=1}^n A_i\right) = \frac{1}{\lambda}\left(\prod_{k=1}^n \left(\prod_{1 \leq i_1 < \cdots < i_k \leq n} (1 + \lambda c_{\lambda,k})\right)^{(-1)^{k-1}} - 1\right), \tag{2.74}$$

where $c_{\lambda,k} = Q_\lambda\left(A_{i_1} \cap \cdots \cap A_{i_k}\right)$.

Proof See the proof of Theorem 2.1. □

Later, we will use the following proposition.

Proposition 2.17 *If X is a finite set, Q_λ is a λ-additive measure on X, $\lambda \in (-1, \infty)$, $\lambda \neq 0$, $A_1, \ldots, A_n \in \mathcal{P}(X)$ and $n \geq 2$, then*

$$Q_\lambda \left(\bigcap_{i=1}^{n} A_i \right) = \frac{1}{\lambda} \left(\prod_{k=1}^{n} \left(\prod_{1 \leq i_1 < \cdots < i_k \leq n} (1 + \lambda d_{\lambda,k}) \right)^{(-1)^{k-1}} - 1 \right), \tag{2.75}$$

where $d_{\lambda,k} = Q_\lambda \left(A_{i_1} \cup \cdots \cup A_{i_k} \right)$.

Proof Let X be a finite set, Q_λ be a λ-additive measure on X, $\lambda \in (-1, \infty)$, $\lambda \neq 0$, $A_1, \ldots, A_n \in \mathcal{P}(X)$, $n \geq 2$. Then, based on Proposition 2.12, the dual measure $Q_{\lambda'}$ of Q_λ, which is given by

$$Q_{\lambda'}(A) = 1 - Q_\lambda(\overline{A}) \tag{2.76}$$

for any $A \in \mathcal{P}(X)$, is also a λ-additive measure with the parameter λ', where

$$\lambda' = -\frac{\lambda}{1 + \lambda}. \tag{2.77}$$

Hence, from Eq. (2.74), we have

$$\begin{aligned} &Q_{\lambda'} \left(\bigcup_{i=1}^{n} A_i \right) = \\ &= \frac{1}{\lambda'} \left(\prod_{k=1}^{n} \left(\prod_{1 \leq i_1 < \cdots < i_k \leq n} \left(1 + \lambda Q_{\lambda'} \left(A_{i_1} \cap \cdots \cap A_{i_k} \right)\right) \right)^{(-1)^{k-1}} - 1 \right). \end{aligned} \tag{2.78}$$

Next, since $A_1, \ldots, A_n \in \mathcal{P}(X)$, $\overline{A}_1, \ldots, \overline{A}_n \in \mathcal{P}(X)$ as well. Now, by applying Eq. (2.78) to the sets $\overline{A}_1, \ldots, \overline{A}$ and by noting Eqs. (2.77) and (2.76), we get

$$\begin{aligned} &1 - Q_\lambda \left(\overline{\overline{A}_1 \cup \cdots \cup \overline{A}_n} \right) = \\ &= -\frac{1 + \lambda}{\lambda} \left(\prod_{k=1}^{n} \left(\prod_{1 \leq i_1 < \cdots < i_k \leq n} \frac{1 + \lambda Q_\lambda \left(\overline{\overline{A}_{i_1} \cap \cdots \cap \overline{A}_{i_k}} \right)}{1 + \lambda} \right)^{(-1)^{k-1}} - 1 \right). \end{aligned}$$

And by utilizing the De Morgan law, from the last equation we get

$$Q_\lambda\left(\bigcap_{i=1}^{n} A_i\right) =$$
$$= \frac{1}{\lambda}\left((1+\lambda)\prod_{k=1}^{n}\left(\prod_{1\le i_1<\cdots<i_k\le n}\frac{1+\lambda d_{\lambda,k}}{1+\lambda}\right)^{(-1)^{k-1}} - 1\right). \tag{2.79}$$

Next, Eq. (2.79) can be written as

$$Q_\lambda\left(\bigcap_{i=1}^{n} A_i\right) =$$
$$= \frac{1}{\lambda}\left(\frac{1+\lambda}{(1+\lambda)^{\sum_{l=1}^{n}(-1)^{l-1}\binom{n}{l}}}\prod_{k=1}^{n}\left(\prod_{1\le i_1<\cdots<i_k\le n}(1+\lambda d_{\lambda,k})\right)^{(-1)^{k-1}} - 1\right),$$

and by utilizing the fact that

$$\sum_{l=1}^{n}(-1)^{l-1}\binom{n}{l} = 1,$$

we get that

$$Q_\lambda\left(\bigcap_{i=1}^{n} A_i\right) = \frac{1}{\lambda}\left(\prod_{k=1}^{n}\left(\prod_{1\le i_1<\cdots<i_k\le n}(1+\lambda d_{\lambda,k})\right)^{(-1)^{k-1}} - 1\right).$$

□

Noting the definitions of the belief- and plausibility measures and the results of Propositions 2.16 and 2.17, we can state the following two propositions.

Proposition 2.18 *Let X be a finite set, Q_λ be a λ-additive measure on X, $\lambda \in (-1, \infty)$ and $\lambda \neq 0$. Then, Q_λ is a belief measure on X if and only if*

$$\frac{1}{\lambda}\left(\prod_{k=1}^{n}\left(\prod_{1\le i_1<\cdots<i_k\le n}(1+\lambda c_{\lambda,k})\right)^{(-1)^{k-1}} - 1\right) > \sum_{k=1}^{n}\sum_{1\le i_1<\cdots<i_k\le n}(-1)^{k-1}c_{\lambda,k}, \tag{2.80}$$

holds for any $A_1, \ldots, A_n \in \mathcal{P}(X)$ and $n \geq 2$, where $c_{\lambda,k} = Q_\lambda\left(A_{i_1} \cap \cdots \cap A_{i_k}\right)$.

Proof Let X be a finite set and Q_λ be a λ-additive measure on X, where $\lambda \in (-1, \infty)$ and $\lambda \neq 0$. Since Q_λ is a λ-additive measure on X, based on Proposition 2.16, the left hand side of Eq. (2.80) is identical to $Q_\lambda(A_1 \cup \cdots \cup A_n)$. And by noting the definition of the belief measures given in Definition 1.4, the statement of this proposition follows. □

Proposition 2.19 *Let X be a finite set, Q_λ be a λ-additive measure on X, $\lambda \in (-1, \infty)$ and $\lambda \neq 0$. Then, Q_λ is a plausibility measure on X if and only if*

$$\frac{1}{\lambda}\left(\prod_{k=1}^{n}\left(\prod_{1 \leq i_1 < \cdots < i_k \leq n}(1+\lambda d_{\lambda,k})\right)^{(-1)^{k-1}} - 1\right) < \sum_{k=1}^{n}\sum_{1 \leq i_1 < \cdots < i_k \leq n}(-1)^{k-1} d_{\lambda,k}, \tag{2.81}$$

holds for any $A_1, \ldots, A_n \in \mathcal{P}(X)$ and $n \geq 2$, where $d_{\lambda,k} = Q_\lambda\left(A_{i_1} \cup \cdots \cup A_{i_k}\right)$.

Proof Let X be a finite set and Q_λ be a λ-additive measure on X, where $\lambda \in (-1, \infty)$ and $\lambda \neq 0$. Noting that Q_λ is a λ-additive measure on X, based on Proposition 2.17, we have that the left hand side of Eq. (2.81) is identical to $Q_\lambda(A_1 \cap \cdots \cap A_n)$. Then, by taking into account the definition of the plausibility measures given in Definition 1.5, the statement of this proposition follows. □

It should be added that the inequalities in Propositions 2.18 and 2.19 may be viewed as new criteria for the belief- and plausibility measures when they are λ-additive measures. Next, noting the results of Propositions 2.18, 2.19 and Theorem 2.6, we can state the following theorem.

Theorem 2.8 *Let X be a finite set, Q_λ be a λ-additive measure on X, $\lambda \in (-1, \infty)$ and $\lambda \neq 0$. Then, the following three statements hold and are equivalent:*

(1a) Q_λ is a belief measure on X if and only if $\lambda > 0$;
(1b) Q_λ is a belief measure on X if and only if the inequality;

$$\frac{1}{\lambda}\left(\prod_{k=1}^{n}\left(\prod_{1 \leq i_1 < \cdots < i_k \leq n}(1+\lambda c_{\lambda,k})\right)^{(-1)^{k-1}} - 1\right) > \sum_{k=1}^{n}\sum_{1 \leq i_1 < \cdots < i_k \leq n}(-1)^{k-1} c_{\lambda,k} \tag{2.82}$$

where $c_{\lambda,k} = Q_\lambda\left(A_{i_1} \cap \cdots \cap A_{i_k}\right)$, holds for any $A_1, \ldots, A_n \in \mathcal{P}(X)$ and $n \geq 2$; and
(1c) the inequality in Eq. (2.82) holds for any $A_1, \ldots, A_n \in \mathcal{P}(X)$ and $n \geq 2$, where $c_{\lambda,k} = Q_\lambda\left(A_{i_1} \cap \cdots \cap A_{i_k}\right)$, if and only if $\lambda > 0$.

Furthermore, the following three statements hold and are equivalent:

(2a) *Q_λ is a plausibility measure on X if and only if $\lambda < 0$;*

(2b) *Q_λ is a plausibility measure on X if and only if the inequality*

$$\frac{1}{\lambda}\left(\prod_{k=1}^{n}\left(\prod_{1\le i_1<\cdots<i_k\le n}(1+\lambda d_{\lambda,k})\right)^{(-1)^{k-1}}-1\right) < \sum_{k=1}^{n}\sum_{1\le i_1<\cdots<i_k\le n}(-1)^{k-1}d_{\lambda,k}, \tag{2.83}$$

where $d_{\lambda,k} = Q_\lambda\left(A_{i_1}\cup\cdots\cup A_{i_k}\right)$, holds for any $A_1,\ldots,A_n \in \mathcal{P}(X)$ and $n \ge 2$; and

(2c) *the inequality in Eq. (2.83) holds for any $A_1,\ldots,A_n \in \mathcal{P}(X)$ and $n \ge 2$, where $d_{\lambda,k} = Q_\lambda\left(A_{i_1}\cup\cdots\cup A_{i_k}\right)$, if and only if $\lambda < 0$.*

Proof By making use of Proposition 2.18 and Theorem 2.6, we immediately see that (1a), (1b) and (1c) are equivalent. Similarly, from Proposition 2.19 and Theorem 2.6 we readily find that (2a), (2b) and (2c) are equivalent. □

2.4.2 The λ-Additive Measure as an Upper and Lower Probability Measure

The following propositions allow us to approximate a probability measure using λ-additive measures.

Proposition 2.20 *Let Pr be a probability measure on the finite set X. Then, for any $A \in \mathcal{P}(X)$*

(1) *if $\lambda > 0$, then*

$$\frac{1}{\lambda}\left((1+\lambda)^{Pr(A)}-1\right) \le Pr(A)$$

(2) *if $-1 < \lambda < 0$, then*

$$\frac{1}{\lambda}\left((1+\lambda)^{Pr(A)}-1\right) \ge Pr(A)$$

(3)

$$\lim_{\lambda\to 0}\left(\frac{1}{\lambda}\left((1+\lambda)^{Pr(A)}-1\right)\right) = Pr(A).$$

Proof Noting the fact that $Pr(A) \in [0,1]$ and $\lambda > -1$, by making use of Bernoulli's inequality, we get the inequality

$$(1+\lambda)^{Pr(A)} \leq 1 + \lambda Pr(A),$$

from which (1) and (2) immediately follows. Next, taking into account the continuity of function h_λ given in Eq. (2.33) (see Eq. (2.36)) we have

$$\lim_{\lambda \to 0} \left(\frac{1}{\lambda} \left((1+\lambda)^{Pr(A)} - 1 \right) \right) = Pr(A).$$

□

Notice that since Pr is a probability measure on the finite set X, based on Theorem 2.3, the measure $Q_\lambda \colon \mathcal{P}(X) \to [0, 1]$, which is given by

$$Q_\lambda(A) = \frac{1}{\lambda} \left((1+\lambda)^{Pr(A)} - 1 \right)$$

for any $A \in \mathcal{P}(X)$, $\lambda \neq 0$, is a λ-additive measure on X. Moreover, based on Theorem 2.6, if $\lambda > 0$, then Q_λ is a belief measure on X, and if $\lambda < 0$, then Q_λ is a plausibility measure on X.

Theorem 2.9 *Let Pr be a probability measure on the finite set X. Furthermore, let $\lambda_1 > 0$ and let λ_2 be given by*

$$\lambda_2 = -\frac{\lambda_1}{1+\lambda_1}.$$

Then, if the set functions $Bl, Pl \colon \mathcal{P}(X) \to [0, 1]$ are given by

$$Bl(A) = \frac{1}{\lambda_1} \left((1+\lambda_1)^{Pr(A)} - 1 \right)$$

and

$$Pl(A) = \frac{1}{\lambda_2} \left((1+\lambda_2)^{Pr(A)} - 1 \right)$$

for any $A \in \mathcal{P}(X)$, then Bl and Pl are a dual pair of belief- and plausibility measures on X, and

$$Bl(A) \leq Pr(A) \leq Pl(A)$$

holds for any $A \in \mathcal{P}(X)$.

Proof Based on Theorem 2.3, Bl and Pl are λ-additive measures on X. Next, by nothing Theorem 2.7, Bl and Pl are a dual pair of belief- and plausibility measures. Now, taking into account the fact that $\lambda_1 > 0$ and so $-1 < \lambda_2 < 0$, Proposition 2.20 implies that the inequality $Bl(A) \leq Pr(A) \leq Pl(A)$ holds for any $A \in \mathcal{P}(X)$. □

2.5 The ν-Additive Measure

It should be added here that the value of parameter λ of a λ-additive measure lies in the interval $(-1, \infty)$. Since this interval is unbounded (from the right hand side) and the zero value of λ does not divide it into two symmetric domains, it is difficult to judge the 'plausibilitiness' or 'beliefness' of a λ-additive measure based on the value of parameter λ. Now, we will demonstrate how the application of an alternatively parameterized λ-additive measure, which we will call the ν-additive measure, can be utilized to characterize the 'plausibilitiness' or 'beliefness' on a normalized scale. We will also show how the application of the ν-additive measure can simplify the numerical solution of Eq. (2.68).

Now, let $Q_\lambda(A)$ be a λ-additive measure and let us assume that $0 \le Q_\lambda(A) < 1$. Then, by using Eq. (2.3), the Q_λ measure of the complement set $\overline{A}$ can be written as

$$Q_\lambda(\overline{A}) = \frac{1 - Q_\lambda(A)}{1 + \lambda Q_\lambda(A)} = \frac{1}{1 + (1+\lambda)\frac{Q_\lambda(A)}{1-Q_\lambda(A)}}. \tag{2.84}$$

In continuous-valued logic, the Dombi form of negation with the neutral value $\nu \in (0, 1)$ is given by the operator $n_\nu \colon [0, 1] \to [0, 1]$ as follows:

$$n_\nu(x) = \begin{cases} \frac{1}{1+\left(\frac{1-\nu}{\nu}\right)^2 \frac{x}{1-x}} & \text{if} x \in [0, 1) \\ 0 & \text{if } x = 1, \end{cases} \tag{2.85}$$

where $x \in [0, 1]$ is a continuous-valued logic variable [22]. Note that the Dombi form of negation is the unique Sugeno's negation [23] with the fix point $\nu \in (0, 1)$. Also, for $Q_\lambda(A) \in [0, 1)$, the formula of λ-additive measure of $Q_\lambda(\overline{A})$ in Eq. (2.84) is the same as the formula of the Dombi form of negation in Eq. (2.85) with $x = Q_\lambda(A)$ and

$$\left(\frac{1-\nu}{\nu}\right)^2 = 1 + \lambda.$$

Based on the definition of λ-additive measures, $\lambda > -1$, and since

$$\lambda = \left(\frac{1-\nu}{\nu}\right)^2 - 1$$

is a bijection between $(0, 1)$ and $(-1, \infty)$, the λ-additive measure of the complement set $\overline{A}$ can be alternatively redefined as

$$Q_\lambda(\overline{A}) = \begin{cases} \frac{1}{1+\left(\frac{1-\nu}{\nu}\right)^2 \frac{Q_\lambda(A)}{1-Q_\lambda(A)}} & \text{if } Q_\lambda(A) \in [0, 1) \\ 0 & \text{if } Q_\lambda(A) = 1, \end{cases} \tag{2.86}$$

where $\left(\frac{1-\nu}{\nu}\right)^2 = 1 + \lambda$, $\nu \in (0, 1)$.

Following this line of thinking, here, we will introduce the ν-additive measure and state some of its properties.

Definition 2.9 The function $Q_\nu \colon \mathcal{P}(X) \to [0, 1]$ is a ν-additive measure on the finite set X, if and only if Q_ν satisfies the following requirements:

(1) $Q_\nu(X) = 1$
(2) for any $A, B \in \mathcal{P}(X)$ and $A \cap B = \emptyset$,

$$Q_\nu(A \cup B) = Q_\nu(A) + Q_\nu(B) + \left(\left(\frac{1-\nu}{\nu}\right)^2 - 1\right) Q_\nu(A) Q_\nu(B), \tag{2.87}$$

where $\nu \in (0, 1)$.

Note that if X is an infinite set, then the continuity of function Q_ν is also required. Here, we state a key proposition that we will frequently utilize later on.

Proposition 2.21 *Let X be a finite set, and let Q_λ and Q_ν be a λ-additive and a ν-additive measure on X, respectively. Then,*

$$Q_\lambda(A) = Q_\nu(A) \tag{2.88}$$

for any $A \in \mathcal{P}(X)$, if and only if

$$\lambda = \left(\frac{1-\nu}{\nu}\right)^2 - 1, \tag{2.89}$$

where $\lambda > -1$, $\nu \in (0, 1)$.

Proof This proposition immediately follows from the definitions of the λ-additive measure and ν-additive measure. □

If Q_ν is a ν-additive measure on the finite set X, then, by utilizing Eq. (2.86), the Q_ν measure of the complement set $\overline{A}$ is

$$Q_\nu(\overline{A}) = \begin{cases} \frac{1}{1+\left(\frac{1-\nu}{\nu}\right)^2 \frac{Q_\nu(A)}{1-Q_\nu(A)}} & \text{if } Q_\nu(A) \in [0, 1) \\ 0 & \text{if } Q_\nu(A) = 1. \end{cases} \tag{2.90}$$

Moreover, as the ν parameter is the neutral value of the Dombi negation operator (see Eq. (2.85)), the following property of the ν-additive measure holds as well.

Proposition 2.22 *Let X be a finite set, Q_ν a ν-additive measure on X and let the set A_ν be given as*

$$A_\nu = \{A \in \mathcal{P}(X) | Q_\nu(A) = \nu\},$$

where $\nu \in (0, 1)$. Then for any $A \in A_\nu$ the Q_ν measure of the complement set $\overline{A}$ is equal to ν; that is, $Q_\nu(\overline{A}) = \nu$.

Proof If $A \in A_\nu$, then $Q_\nu(A) = \nu$ and utilizing the ν-additive negation given by Eq. (2.90), we have

$$Q_\nu(\overline{A}) = \frac{1}{1 + \left(\frac{1-\nu}{\nu}\right)^2 \frac{\nu}{1-\nu}} = \nu.$$

□

This result means that the ν-additive complement operation may be viewed as a complement operation characterized by its fix point ν. Notice that the value of parameter ν of a ν-additive measure is in the bounded interval (0, 1), while the value of parameter λ of the corresponding λ-additive measure is in the unbounded interval $(-1, \infty)$. It means that the parameter ν characterizes the 'plausibilitiness' or 'beliefness' of the ν-additive measure on a normalized scale. Moreover, $\nu = 0.5$, which corresponds to a probability measure, divides the interval (0, 1) symmetrically into the parameter domains of belief- ($\nu \in (0, 0.5)$) and plausibility ($\nu \in (0.5, 1)$) measures. It should be also added that when we seek to numerically solve Eq. (2.68), then we need to search for the solution in the interval $(-1, \infty)$. However, if we utilize the corresponding ν-additive measure, then we need to search for the value of ν in the interval (0, 1), which considerably simplifies the numerical computation.

2.6 Properties of the ν-Additive Measure

In this section, we will discuss the main properties of the ν-additive measure. Here, Q_ν will always denote a ν-additive measure with the parameter $\nu \in (0, 1)$.

2.6.1 Dual ν-Additive Measures and Their Properties

Later, we will utilize the concept of the dual pair of ν-additive measures.

Definition 2.10 Let Q_{ν_1} and Q_{ν_2} be two ν-additive measures on the finite set X. Then, Q_{ν_1} and Q_{ν_2} are said to be a dual pair of ν-additive measures if and only if

$$Q_{\nu_1}(A) + Q_{\nu_2}(\overline{A}) = 1$$

holds for any $A \in \mathcal{P}(X)$, where $\nu_1, \nu_2 \in (0, 1)$.

Here, we will demonstrate some key properties of the ν-additive measure related to a dual pair of ν-additive measures. Theorem 2.5 can be stated in terms of the ν-additive measure as follows.

Theorem 2.10 *Let Q_{ν_1} and Q_{ν_2} be two ν-additive measures on the finite set X. Then, Q_{ν_1} and Q_{ν_2} are a dual pair of ν-additive measures if and only if*

$$\nu_1 + \nu_2 = 1,$$

where $\nu_1, \nu_2 \in (0, 1)$.

Proof Utilizing Proposition 2.21, this theorem immediately follows from Theorem 2.5. □

Utilizing the definition of the dual pair of ν-additive measures, Corollary 2.1 can be stated in terms of the ν-additive measure as follows.

Corollary 2.2 *Let Q_{ν_1} and Q_{ν_2} be a dual pair of ν-additive measures on the finite set X. Then, $\nu_1 \in [1/2, 1)$ if and only if $\nu_2 \in (0, 1/2]$.*

Proof Taking into account Proposition 2.21, this corollary immediately follows from Corollary 2.1. □

It should be mentioned here that one of the λ parameter values of a dual pair of λ-additive measures is always in the unbounded interval $[0, \infty)$. At the same time, the ν parameters of a dual pair of ν-additive measures are both in a bounded interval; namely, one of them is in the interval $(0, 1/2]$ and the other one is in the interval $[1/2, 1)$.

2.6.1.1 Connection with Belief-, Probability- and Plausibility Measures

Here, we will discuss some important properties of the ν-additive measure and show how it is connected to the belief-, probability- and plausibility measures.

Note that in terms of the ν-additive measure, Theorem 2.6 can be stated as follows.

Theorem 2.11 *Let X be a finite set and let Q_ν be a ν-additive measure on X. Then, on set X, Q_ν is a*

(1) belief measure if and only if $0 < \nu \leq 1/2$
(2) probability measure if and only if $\nu = 1/2$
(3) plausibility measure if and only if $1/2 \leq \nu < 1$.

Proof Taking into account Proposition 2.21, this theorem immediately follows from Theorem 2.6. □

Figure 2.1 shows the connection between $Q_\nu(\overline{A})$ and $Q_\nu(A)$ for various values of parameter ν of the ν-additive measure Q_ν. From this figure, in accordance with Theorem 2.11, we notice the following. If $\nu = 1/2$, then Q_ν is a probability measure and so $Q_\nu(\overline{A}) = 1 - Q_\nu(A)$. If $0 < \nu \leq 1/2$, then Q_ν is a belief measure and $Q_\nu(\overline{A}) \leq 1 - Q_\nu(A)$. If $1/2 \leq \nu < 1$, then Q_ν is a plausibility measure and

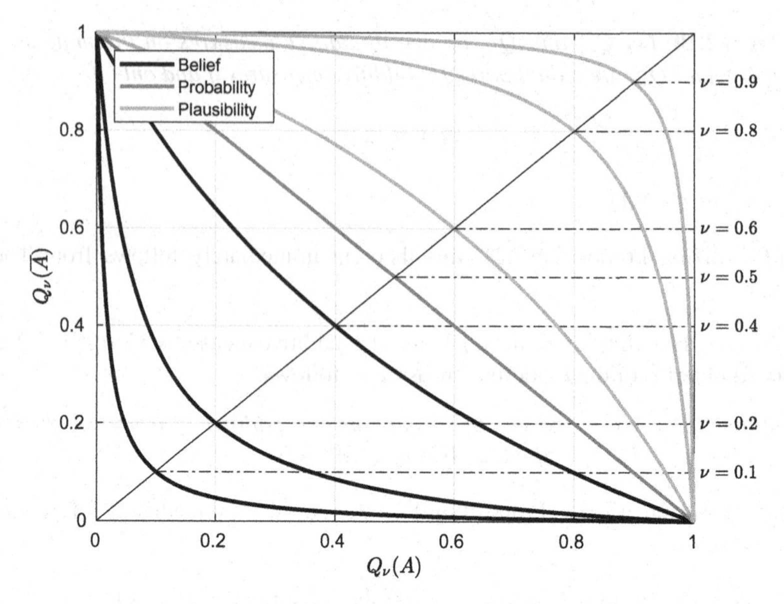

Fig. 2.1 ν-additive measures of set A versus ν-additive measures of complement of A

$Q_\nu(\overline{A}) \geq 1 - Q_\nu(A)$. Moreover, in accordance with Eq. (2.90), for a given set A, $Q_\nu(\overline{A})$ increases with the value of parameter ν. That is, the smaller the value of parameter ν, the stronger the complement operation is. It also means that any belief measure of a complement set is always less than or equal to any plausibility measure of the same complement set.

Proposition 2.14 can be stated in terms of the ν-additive measure as follows.

Proposition 2.23 *Let Q_{ν_1} and Q_{ν_2} be two ν-additive measures on the finite set X. Then, Q_{ν_1} and Q_{ν_2} are a dual pair of belief- and plausibility measures on X if and only if they are a dual pair of ν-additive measures on X.*

Proof Taking into account Proposition 2.21, this proposition directly follows from Proposition 2.14. □

Theorem 2.7 can be stated in terms of the ν-additive measure as follows.

Theorem 2.12 *Let Q_{ν_1} and Q_{ν_2} be two ν-additive measures on the finite set X. Then, Q_{ν_1} and Q_{ν_2} are a dual pair of belief- and plausibility measures on X if and only if*

$$\nu_1 + \nu_2 = 1,$$

where $\nu_1, \nu_2 \in (0, 1)$.

Proof Based on Proposition 2.21, this theorem immediately follows from Theorem 2.7. □

It should be added here that a ν-additive measure may be supermodular or submodular depending on the value of its parameter ν.

Definition 2.11 The set function $f\colon \mathcal{P}(X) \to \mathbb{R}$ on the finite set X is said to be submodular if

$$f(A) + f(B) \geq f(A \cup B) + f(A \cap B)$$

holds for any $A, B \in \mathcal{P}(X)$.

Definition 2.12 The set function $f\colon \mathcal{P}(X) \to \mathbb{R}$ on the finite set X is said to be supermodular if

$$f(A) + f(B) \leq f(A \cup B) + f(A \cap B)$$

holds for any $A, B \in \mathcal{P}(X)$.

Corollary 2.3 *A ν-additive measure is supermodular if $\nu \in (0, 1/2]$, and it is submodular if $\nu \in [1/2, 1)$.*

Proof Since every belief measure is supermodular and every plausibility measure is submodular, this corollary immediately follows from Theorem 2.11. □

2.7 Inequalities for ν-Additive Measures

2.7.1 Characteristic Inequalities for Belief- and Plausibility Measures Using the ν-Additive Measure

Noting the definition of ν-additive measures and the result of Proposition 2.21, Theorem 2.8 can be stated for ν-additive measures as follows.

Theorem 2.13 *Let X be a finite set, Q_ν be a ν-additive measure on X, $\nu \in (0, 1)$ and $\nu \neq 1/2$. Then, the following three statements hold and are equivalent:*

(1a) Q_ν is a belief measure on X if and only if $\nu \in (0, 1/2)$;
(1b) Q_ν is a belief measure on X if and only if the inequality

$$\frac{\nu^2}{1-2\nu}\left(\prod_{k=1}^{n}\left(\prod_{1\leq i_1<\cdots<i_k\leq n}\left(1+\frac{1-2\nu}{\nu^2}c_{\nu,k}\right)\right)^{(-1)^{k-1}}-1\right) > \sum_{k=1}^{n}\sum_{1\leq i_1<\cdots<i_k\leq n}(-1)^{k-1}c_{\nu,k}, \tag{2.91}$$

where $c_{\nu,k} = Q_\nu\left(A_{i_1} \cap \cdots \cap A_{i_k}\right)$, holds for any $A_1, \ldots, A_n \in \mathcal{P}(X)$ and $n \geq 2$; and

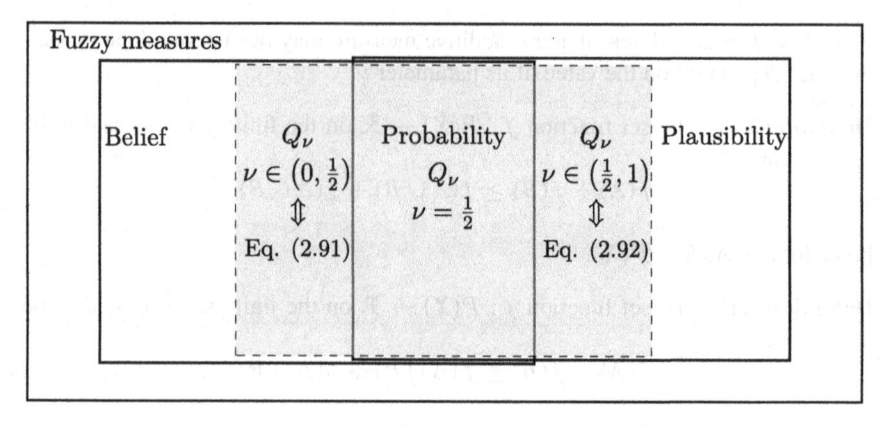

Fig. 2.2 Relations among the belief-, probability-, plausibility- and ν-additive measures

(1c) *the inequality in Eq.* (2.91) *holds for any* $A_1, \ldots, A_n \in \mathcal{P}(X)$ *and* $n \geq 2$, *where* $c_{\nu,k} = Q_\nu\left(A_{i_1} \cap \cdots \cap A_{i_k}\right)$, *if and only if* $\nu \in (0, 1/2)$.

Furthermore, the following three statements all hold and are equivalent:

(2a) Q_ν *is a plausibility measure on* X *if and only if* $\nu \in (1/2, 1)$;
(2b) Q_ν *is a plausibility measure on* X *if and only if the inequality*

$$\frac{\nu^2}{1-2\nu}\left(\prod_{k=1}^{n}\left(\prod_{1\leq i_1<\cdots<i_k\leq n}\left(1+\frac{1-2\nu}{\nu^2}d_{\nu,k}\right)\right)^{(-1)^{k-1}}-1\right) < \sum_{k=1}^{n}\sum_{1\leq i_1<\cdots<i_k\leq n}(-1)^{k-1}d_{\nu,k}, \tag{2.92}$$

where $d_{\nu,k} = Q_\nu\left(A_{i_1} \cup \cdots \cup A_{i_k}\right)$, *holds for any* $A_1, \ldots, A_n \in \mathcal{P}(X)$ *and* $n \geq$ 2; *and*

(2c) *the inequality in Eq.* (2.92) *holds for any* $A_1, \ldots, A_n \in \mathcal{P}(X)$ *and* $n \geq 2$, *where* $d_{\nu,k} = Q_\nu\left(A_{i_1} \cup \cdots \cup A_{i_k}\right)$, *if and only if* $\nu \in (1/2, 1)$.

Proof By noting Proposition 2.21, the theorem immediately follows from Theorem 2.8. □

Notice that the inequalities given in Eqs. (2.91) and (2.92) may be interpreted as new criteria for the belief- and plausibility measures when they are ν-additive measures. Figure 2.2 gives an overview of the relations among the belief-, probability-, plausibility- and ν-additive measures. The gray-colored area in Fig. 2.2 corresponds to the ν-additive measures[1].

[1]Reprinted from József Dombi and Tamás Jónás, Inequalities for λ-additive measures based on the application of the general Poincaré formula for λ-additive measures, Fuzzy Sets and Systems, https://doi.org/10.1016/j.fss.2019.09.007, Copyright (2019), with permission from Elsevier.

2.7.2 The ν-Additive Measure as an Upper and Lower Probability Measure

Utilizing the definition of the ν-additive measure, Theorem 2.3 can be stated as follows.

Theorem 2.14 *Let the function* $h_\nu \colon [0, 1] \to [0, 1]$ *be given by*

$$h_\nu(x) = \begin{cases} \dfrac{\left(\left(\frac{1-\nu}{\nu}\right)^2\right)^x - 1}{\left(\frac{1-\nu}{\nu}\right)^2 - 1}, & \text{if } \nu \neq \frac{1}{2} \\ x, & \text{if } \nu = \frac{1}{2}, \end{cases}$$

where $\nu \in (0, 1)$. *Furthermore, let* Σ *be a* σ*-algebra over the set* X *and let* Q_ν *and* Pr *be two continuous functions on the space* (X, Σ) *such that*

$$Q_\nu = h_\nu \circ Pr. \tag{2.93}$$

Then, Pr *is a probability measure on* (X, Σ) *if and only if* Q_ν *is a* ν*-additive measure on* (X, Σ).

Proof Taking into account Proposition 2.21, this theorem immediately follows from Theorem 2.3. □

Based on the result of Theorem 2.14, the formula in Eq. (2.93) may be viewed as a transformation between probability measures and ν-additive measures.

Proposition 2.20 can alternatively be stated as follows.

Proposition 2.24 *Let* Pr *be a probability measure on the finite set* X. *Then, for any* $A \in \mathcal{P}(X)$

(1) if $\nu \in (0, 1/2)$, *then*

$$\frac{\nu^2}{1-2\nu}\left(\left(\frac{1-\nu}{\nu}\right)^{2Pr(A)} - 1\right) \leq Pr(A)$$

(2) if $\nu \in (1/2, 1)$, *then*

$$\frac{\nu^2}{1-2\nu}\left(\left(\frac{1-\nu}{\nu}\right)^{2Pr(A)} - 1\right) \geq Pr(A)$$

(3)

$$\lim_{\nu \to \frac{1}{2}} \frac{\nu^2}{1-2\nu}\left(\left(\frac{1-\nu}{\nu}\right)^{2Pr(A)} - 1\right) = Pr(A).$$

Proof By applying the substitution $\lambda = \left(\frac{1-\nu}{\nu}\right)^2 - 1$, where $\lambda > -1$ and $\nu \in (0, 1)$, this proposition readily follows from Proposition 2.20. □

It should be added that since Pr is a probability measure on the finite set X, based on Theorem 2.14, the measure $Q_\nu : \mathcal{P}(X) \to [0, 1]$, which is given by

$$Q_\nu(A) = \frac{\nu^2}{1-2\nu}\left(\left(\frac{1-\nu}{\nu}\right)^{2Pr(A)} - 1\right)$$

for any $A \in \mathcal{P}(X)$, $\nu \neq \frac{1}{2}$, is a ν-additive measure on X. Moreover, the next theorem holds.

Theorem 2.15 *Let Pr be a probability measure on the finite set X. Furthermore, let $\nu_1 \in (0, 1/2)$ and*

$$\nu_1 + \nu_2 = 1.$$

Then, if the set functions $Bl, Pl : \mathcal{P}(X) \to [0, 1]$ are given by

$$Bl(A) = \frac{\nu_1^2}{1-2\nu_1}\left(\left(\frac{1-\nu_1}{\nu_1}\right)^{2Pr(A)} - 1\right)$$

and

$$Pl(A) = \frac{\nu_2^2}{1-2\nu_2}\left(\left(\frac{1-\nu_2}{\nu_2}\right)^{2Pr(A)} - 1\right)$$

for any $A \in \mathcal{P}(X)$, then Bl and Pl are a dual pair of belief- and plausibility measures on X, and

$$Bl(A) \leq Pr(A) \leq Pl(A)$$

holds for any $A \in \mathcal{P}(X)$.

Proof Based on Theorem 2.14, Bl and Pl are ν-additive measures on X. Next, by nothing Theorem 2.12, Bl and Pl are a dual pair of belief- and plausibility measures. Now, taking into account the fact that $\nu_1 \in (0, 0.5)$ and so $\nu_2 \in (0.5, 1)$, Proposition 2.24 implies that the inequality $Bl(A) \leq Pr(A) \leq Pl(A)$ holds for any $A \in \mathcal{P}(X)$. □

2.8 Connections of ν-Additive (λ-Additive) Measures with Other Areas

2.8.1 Connection with Rough Sets

It is a well-known fact that the belief- and plausibility measures are connected with the rough set theory (see [24–26]). Here, we will show how the ν-additive (λ-additive) measures are connected with the rough set theory.

Definition 2.13 Let X be a finite set, and let $R \subseteq X \times X$ be a binary equivalence relation on X. The pair $(\underline{R}(A), \overline{R}(A))$ is said to be the rough set of $A \subseteq X$ in the approximation space (X, R) if

$$\underline{R}(A) = \{x \in X | [x]_R \subseteq A\}$$
$$\overline{R}(A) = \{x \in X | [x]_R \cap A \neq \emptyset\},$$

where $[x]_R$ is the R-equivalence class containing x.

The concept of a rough set was introduced by Pawlak [27]. The rough set $(\underline{R}(A), \overline{R}(A))$ can be utilized to characterize the set A by the pair of lower and upper approximations $(\underline{R}(A), \overline{R}(A))$. The lower approximation $\underline{R}(A)$ is the union of all elementary sets that are subsets of A, and the upper approximation $\overline{R}(A)$ is the union of all elementary sets that have a non-empty intersection with A. Note that the definitions of $\underline{R}(A)$ and $\overline{R}(A)$ are equivalent to the following statement: an element of X necessarily belongs to A if all of its equivalent elements belong to A, while an element of X possibly belongs to A if at least one of its equivalent elements belongs to A [26] (see Fig. 2.3).

Let the functions $\underline{q}, \overline{q} : \mathcal{P}(X) \to [0, 1]$ be defined as follows:

$$\underline{q}(A) = \frac{|\underline{R}(A)|}{|X|}, \ \overline{q}(A) = \frac{|\overline{R}(A)|}{|X|}$$

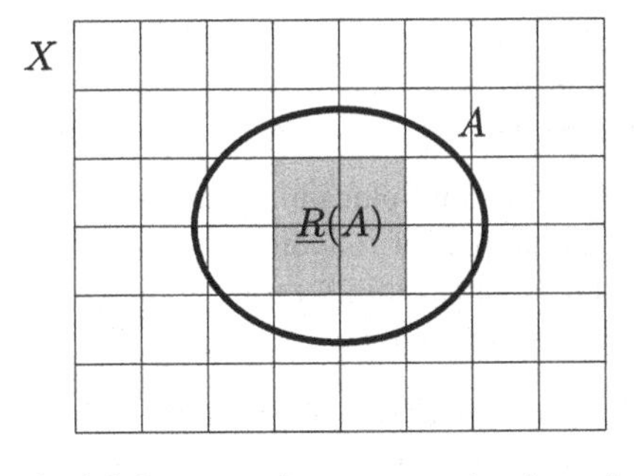

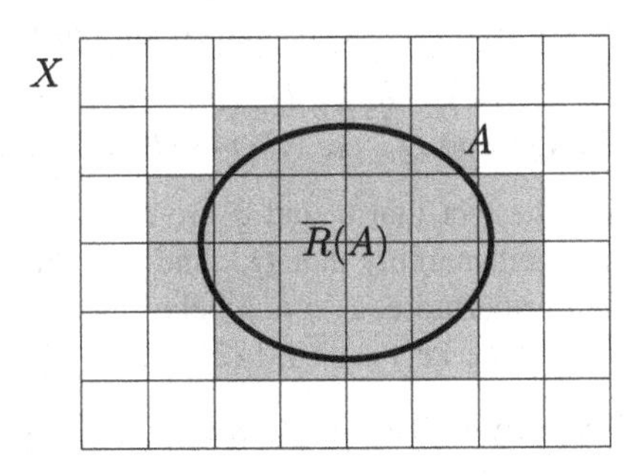

Fig. 2.3 Lower- and upper approximations of set A

for any $A \subseteq X$. Skowron [28, 29] showed that the functions $\underline{q}$ and $\overline{q}$ are a dual pair of belief- and plausibility measures and the corresponding basic probability assignment is $m(A^*) = |A^*|/|X|$ for all $A^* \in X/R$, and 0 otherwise. Furthermore, Yao and Lingras [25] demonstrated that if Pl and Bl are a dual pair of plausibility and belief functions on X and m is the basic probability assignment of Bl satisfying the conditions: (1) the set of focal elements of m is a partition of X, (2) $m(A^*) = |A^*|/|X|$ for every focal element A^* of m, then there exists an equivalence relation R on the set X, such that the induced qualities of upper and lower approximations satisfy

$$\underline{q}(A) = Bl(A),\ \overline{q}(A) = Pl(A)$$

for any $A \subseteq X$ [26].

Based on these results and on our proposition findings, we will establish some connections between rough sets and ν-additive measures by using the following propositions.

Proposition 2.25 *Let Q_{ν_1} and Q_{ν_2} be two ν-additive measures on the finite set X, and let $R \subseteq X \times X$ be a binary equivalence relation on X. Furthermore, let $(\underline{R}(A), \overline{R}(A))$ be the rough set of $A \in \mathcal{P}(X)$ with respect to the approximation space (X, R) and let the functions $\underline{q}, \overline{q} \colon \mathcal{P}(X) \to [0, 1]$ be given by*

$$\underline{q}(A) = \frac{|\underline{R}(A)|}{|X|},\ \overline{q}(A) = \frac{|\overline{R}(A)|}{|X|},$$

where $\underline{R}(A)$ and $\overline{R}(A)$ are the lower- and upper approximations of A, respectively, for any $A \in \mathcal{P}(X)$. If the equations

$$Q_{\nu_1}(A) = \underline{q}(A),\ Q_{\nu_2}(A) = \overline{q}(A), \tag{2.94}$$

hold for any $A \in \mathcal{P}(X)$, then Q_{ν_1} and Q_{ν_2} are a dual pair of ν-additive measures on X with $\nu_1 \in (0, 1/2]$, $\nu_2 \in [1/2, 1)$.

Proof Based on Skowron's results in [28, 29], if the conditions of this proposition are satisfied, then the functions $\underline{q}$ and $\overline{q}$ are a dual pair of belief- and plausibility measures on X. Hence, the conditions that

(i) $Q_{\nu_1}(A) = \underline{q}(A)$, $Q_{\nu_2}(A) = \overline{q}(A)$ hold for any $A \in \mathcal{P}(X)$
(ii) Q_{ν_1} and Q_{ν_2} are two ν-additive measures on X

and the fact that $\underline{q}$ and $\overline{q}$ are a dual pair of belief- and plausibility measures on X together imply that Q_{ν_1} and Q_{ν_2} are also a dual pair of ν-additive measures on X. Furthermore, as $\underline{q}$ is a belief measure and $\overline{q}$ is a plausibility measure, based on Theorem 2.11, $\nu_1 \in (0, 1/2]$ and $\nu_2 \in [1/2, 1)$ hold as well. □

Proposition 2.26 *If Q_{ν_1} and Q_{ν_2} are a dual pair of ν-additive measures on the finite set X with $\nu_1 \in (0, 1/2]$, $\nu_2 \in [1/2, 1)$ and m is a basic probability assignment that satisfies the conditions:*

(1) The set of focal elements of m *is a partition of* X
(2) $m(A^*) = |A^*|/|X|$ *for every focal element* A^* *of* m
(3) $m(A^*) = \sum_{B \subseteq A^*} (-1)^{|A^*|-|B|} Q_{\nu_1}(B)$ *for any* $A^* \in \mathcal{P}(X)$,

then there exists an equivalence relation R *on the set* X*, such that the equations*

$$Q_{\nu_1}(A) = \underline{q}(A), \ Q_{\nu_2}(A) = \overline{q}(A)$$

hold for any $A \in \mathcal{P}(X)$*, where* $(\underline{R}(A), \overline{R}(A))$ *is the rough set of* A *with respect to the approximation space* (X, R)*,* $\underline{q}, \overline{q}: \mathcal{P}(X) \to [0, 1]$ *are given as*

$$\underline{q}(A) = \frac{|\underline{R}(A)|}{|X|}, \ \overline{q}(A) = \frac{|\overline{R}(A)|}{|X|},$$

and $\underline{R}(A)$ *and* $\overline{R}(A)$ *are the lower- and upper approximations of* A*, respectively.*

Proof Based on result of Yao and Lingras [25], if Pl and Bl are a dual pair of plausibility and belief functions on X and m is the basic probability assignment of Bl satisfying the conditions: (i) the set of focal elements of m is a partition of X, (ii) $m(A^*) = |A^*|/|X|$ for every focal element A^* of m, then there exists an equivalence relation R on the set X, such that the induced qualities of upper and lower approximations satisfy

$$\underline{q}(A) = Bl(A), \ \overline{q}(A) = Pl(A)$$

for any $A \in \mathcal{P}(X)$. Therefore, it is sufficient to show that if the conditions of our proposition are satisfied, then Q_{ν_1} is a belief measure on X, Q_{ν_2} is a plausibility measure on X, and m is the basic probability assignment of the belief measure Q_{ν_1}.

Let us assume that the conditions of this proposition are satisfied. Then, since Q_{ν_1} and Q_{ν_2} are a dual pair of ν-additive measures on the finite set X, based on Proposition 2.23, Q_{ν_1} and Q_{ν_2} are a dual pair of belief- and plausibility measures on X. Furthermore, as $\nu_1 \in (0, 1/2]$ and $\nu_2 \in [1/2, 1)$, based on Theorem 2.11, Q_{ν_1} is a belief measure on X and Q_{ν_2} is a plausibility measure on X, and so condition (3) means that m is the basic probability assignment of the belief measure Q_{ν_1}. That is, we have shown that if the conditions of this proposition are satisfied, then all the conditions that are required to apply the result of Yao and Lingras [25] are satisfied as well. □

2.8.2 The λ-Additive Measure and the Multi-attribute Utility Function

Here we will state interesting analogies between the λ-additive measure and the multi-attribute utility function. Let $X_1, X_2, \ldots, X_n$ be attributes, where each X_i may be either a scalar attribute or a vector of scalar attributes ($i = 1, 2, \dots, n$). Furthermore, let the consequence space X be a rectangular subset of the n-dimensional Euclidean space. Then a specific consequence may be given by a vector $(x_1, x_2, \dots, x_n)$, where x_i is a particular value of the attribute X_i ($i = 1, 2, \dots, n$). The utility function $u\colon X \to \mathbb{R}$, which is assumed to be continuous, assigns a utility value to the consequence $(x_1, x_2, \dots, x_n)$; that is, the utility of consequence $(x_1, x_2, \dots, x_n)$ is $u(x_1, x_2, \dots, x_n)$ [21]. Here, we will utilize the concept of the utility independence of attributes (see, e.g. [30]).

Definition 2.14 Attribute X_i is utility independent of attribute X_j if conditional preferences for lotteries over X_i given a fixed value for X_j do not depend on the particular value of X_j.

Keeney and Raiffa [30] proved the following proposition which states that the mutual utility independence of attributes implies a multiplicative multi-attribute utility function.

Proposition 2.27 *If $X_1, X_2, \dots, X_n$ are mutually utility independent attributes, then*

$$u_M(x_1, x_2, \dots, x_n) = \frac{1}{k}\left(\prod_{i=1}^{n}(1 + kk_i u_i(x_i)) - 1\right), \tag{2.95}$$

where $u_M\colon \mathbb{R}^n \to [0, 1]$ is a multi-attribute utility function, $u_i : \mathbb{R} \to [0, 1]$ are utility functions, k_i is the weight of attribute X_i with $0 < k_i < 1$, and $k > -1$, $k \neq 0$ is a scaling constant ($i = 1, 2, \dots, n$).

Proof See [30]. □

The multi-attribute utility function u_M in Eq. (2.95) plays a key role in multi-attribute utility theory and it can be written as

$$1 + ku_M(x_1, x_2, \dots, x_n) = \prod_{i=1}^{n}(1 + kk_i u_i(x_i)). \tag{2.96}$$

If k is positive in Eq. (2.96), then $u^*(x_1, x_2, \dots, x_n) = 1 + ku_M(x_1, x_2, \dots, x_n)$ is a multi-attribute utility function, $u_i^*(x_i) = 1 + kk_i u_i(x_i)$ are utility functions and $u^*(x_1, x_2, \dots, x_n) = \prod_{i=1}^{n} u_i^*(x_i)$, where $i = 1, 2, \dots, n$. Similarly, if k is negative in Eq. (2.96), then $u^*(x_1, x_2, \dots, x_n) = -(1 + ku_M(x_1, x_2, \dots, x_n))$ is a multi-attribute utility function, $u_i^*(x_i) = -(1 + kk_i u_i(x_i))$ are utility functions and $-u^*(x_1, x_2, \dots, x_n) = (-1)^n \prod_{i=1}^{n} u_i^*(x_i)$, where $i = 1, 2, \dots, n$. That is, Eq. (2.96) describes a multiplicative relationship between the multi-attribute utility function and the individual univariate utility functions. Hence, Eq. (2.95) is referred to as the multi-attribute multiplicative utility function.

We can see that the right hand side of Eq. (2.31) with $\lambda > -1$, $\lambda \neq 0$ has the same form as the right hand side of Eq. (2.95). It means that there is an interesting connection between the λ-additive measures and the multi-attribute multiplicative utility function. Namely, a λ-additive measure with $\lambda \neq 0$ of the union of n pairwise disjoint sets is computed in the same way as the multi-attribute utility of n mutually utility independent attributes.

Here, the formula in Eq. (2.95) can be written as

$$u_M(x_1, x_2, \ldots, x_n) = \sum_{i=1}^{n} k_i u_i(x_i) + \\ + \sum_{r=2}^{n} k^{r-1} \sum_{\substack{1 \le i_1 < \cdots \\ \cdots < i_r \le n}} k_{i_1} \cdots k_{i_r} u_{i_1}(x_{i_1}) \cdots u_{i_r}(x_{i_r}) \tag{2.97}$$

from which

$$\lim_{k \to 0} u_M(x_1, x_2, \ldots, x_n) = \sum_{i=1}^{n} k_i u_i(x_i).$$

Note that

$$u_A(x_1, x_2, \ldots, x_n) = \sum_{i=1}^{n} k_i u_i(x_i) \tag{2.98}$$

is the so-called multi-attribute additive utility function [21]. We can get Eq. (2.98) from Eq. (2.97) by allowing for $k = 0$.

Definition 2.15 Two attributes X_i and X_j are additive independent if the paired preference comparison of any two lotteries, defined by two joint probability distributions on $X_i \times X_j$, depends only on their marginal distributions.

It can be shown that if and only if the preferences over lotteries on attributes $X_1, X_2, \ldots, X_n$ depend only on their marginal probability distributions (i.e. the attributes are additive independent), then the n-attribute utility function is additive [30].

Notice that the right hand side of Eq. (2.31) with $\lambda = 0$ has the same form as the right hand side of Eq. (2.98). It means that a λ-additive measure with $\lambda = 0$ of the union of n pairwise disjoint sets is computed in the same way as the multi-attribute utility of n additive independent attributes.

Table 2.1 summarizes the analogies between the λ-additive measures and the multi-attribute utility functions.

2.8.3 *The λ-Additive Measure and Some Operators of Continuous-Valued Logic*

Here, we will state a formal connection between the λ-additive measure and certain operators of continuous-valued logic.

Table 2.1 λ-additive measure of union of pairwise disjoint sets and utility value of consequence $(x_1, x_2, \ldots, x_n)$

Q_λ λ-additive measure of $\bigcup_{i=1}^{n} A_i$; $\lambda > -1$	
$\lambda = 0$:	$\sum_{i=1}^{n} Q_\lambda(A_i)$
$\lambda \neq 0$:	$\frac{1}{\lambda}\left(\prod_{i=1}^{n}(1+\lambda Q_\lambda(A_i)) - 1\right)$
Multi-attribute utility $u(x_1, x_2, \ldots, x_n)$; $k > -1$	
$k = 0$:	$\sum_{i=1}^{n} k_i u_i(x_i)$
$k \neq 0$:	$\frac{1}{k}\left(\prod_{i=1}^{n}(1+k k_i u_i(x_i)) - 1\right)$

Definition 2.16 The generalized Dombi operator $o_{GD,\gamma}^{(\alpha)} : [0, 1]^n \to [0, 1]$ is given by

$$o_{GD,\gamma}^{(\alpha)}(\mathbf{x}) =$$

$$= \frac{1}{1 + \left(\frac{1}{\gamma}\left(\prod_{i=1}^{n}\left(1 + \gamma\left(\frac{1-x_i}{x_i}\right)^{\alpha}\right) - 1\right)\right)^{1/\alpha}}, \tag{2.99}$$

where $\mathbf{x} = (x_1, x_2, \ldots, x_n)$, $x_1, x_2, \ldots, x_n$ are continuous-valued logic variables, $\alpha \in (-\infty, \infty)$ and $\gamma \in (0, \infty)$ [22].

It can be shown that if $\alpha > 0$, then $o_{GD,\gamma}^{(\alpha)}$ is a conjunction operator, and if $\alpha < 0$, then $o_{GD,\gamma}^{(\alpha)}$ is a disjunction operator (see [22]). Moreover, the operator $o_{GD,\gamma}^{(\alpha)}$ is general because depending on its parameter values it can cover a range of familiar fuzzy conjunction and disjunction operators including the Dombi operators [31], the product operators [22], the Einstein operators [32], the Hamacher operators [33], the drastic operators [34] and the min-max operators [35]. Table 2.2 summarizes the operators that the generalized Dombi operator class can cover.

Table 2.2 Operators covered by the generalized Dombi operator class

Operator	γ	Conjunction	Disjunction
		Value of α	
Dombi	0	$\alpha > 0$	$\alpha < 0$
Product	1	1	–1
Einstein	2	1	–1
Hamacher	$\gamma \in (0, \infty)$	1	–1
Drastic	∞	$\alpha > 0$	$\alpha < 0$
Min-max	0	∞	$-\infty$

Here, from Eq. (2.99) we have

$$\left(\frac{1-o^{(\alpha)}_{GD,\gamma}(\mathbf{x})}{o^{(\alpha)}_{GD,\gamma}(\mathbf{x})}\right)^{\alpha} = \\ = \frac{1}{\gamma}\left(\prod_{i=1}^{n}\left(1+\gamma\left(\frac{1-x_i}{x_i}\right)^{\alpha}\right)-1\right) \tag{2.100}$$

for any $o^{(\alpha)}_{GD,\gamma}(\mathbf{x}) \in (0, 1]$. Next, the generator function $g : (0, 1] \to [0, \infty)$ of Dombi operators [31] is given by

$$g(x) = \left(\frac{1-x}{x}\right)^{\alpha}.$$

Utilizing this function, Eq. (2.100) can be written as

$$g\left(o^{(\alpha)}_{GD,\gamma}(x_1, x_2, \ldots, x_n)\right) = \frac{1}{\gamma}\left(\prod_{i=1}^{n}(1+\gamma g(x_i)) - 1\right). \tag{2.101}$$

Recall that based on Proposition 2.2, if X is a finite set, Q_λ is a λ-additive measure on X, $\lambda > -1$, $\lambda \neq 0$ and $A_1, A_2, \ldots, A_n \in \mathcal{P}(X)$ are pairwise disjoint sets, then

$$Q_\lambda\left(\bigcup_{i=1}^{n} A_i\right) = \frac{1}{\lambda}\left(\prod_{i=1}^{n}(1+\lambda Q_\lambda(A_i)) - 1\right). \tag{2.102}$$

From Eqs. (2.101) and (2.102), we notice an interesting analogy. Namely, a λ-additive measure with $\lambda \neq 0$ of the union of n pairwise disjoint sets is computed in the same way as the value of the generator function of Dombi operator for the value of the generalized Dombi operation over n continuous-valued logic variables. It should be added that this analogy is just a formal one since $g(x_i) \in (0, \infty)$ and $Q_\lambda(A_i) \in [0, 1]$, and $g(x_i)$ and $Q_\lambda(A_i)$ have different meanings.

2.9 Summary

In this chapter, we revisited the λ-additive measure (Sugeno λ-measure) and summarized its most important properties. Here, we presented the general formula for the λ-additive measure of the union of n sets and demonstrated that the Poincaré formula of probability theory is just a limit case of the general formula for the λ-additive measure of union of n sets. That is, our formula may be viewed as the generalization of the Poincaré formula. We introduced novel inequalities that are connected with the λ-additive-, belief- and plausibility measures. Also, we pointed out that these inequalities, which are based on the application of the general Poincaré formula for λ-additive measures, can be utilized as new criteria for the belief- and plausibility measures when they are λ-additive measures. Next, we introduced the so-called ν-additive measure as an alternatively parameterized λ-additive measure. The motivation for introducing the ν-additive measure lies in the fact that its parameter $\nu \in (0, 1)$ has an important semantic meaning as it is the fix point of the complement operation. Here, by utilizing the ν-additive measure, some well-known results concerning the λ-additive measure were put into a new light and rephrased in more advantageous forms. We discussed here how the ν-additive measure is connected with the belief-, probability- and plausibility measures. Also, we discussed here how the ν-additive measures and λ-additive measures are connected with rough sets, multi-attribute utility functions and with certain operators of fuzzy logic.

References

1. E. Pap, *Null-Additive Set Functions*, vol. 337 (Kluwer Academic Publishers, Dordrecht, 1995)
2. E. Pap, Pseudo-additive measures and their applications, *Handbook of Measure Theory* (Elsevier, Amsterdam, 2002), pp. 1403–1468
3. Z. Wang, G.J. Klir, *Generalized Measure Theory*. IFSR International Series in Systems Science and Systems Engineering (Springer US, New York, 2010)
4. L. Jin, R. Mesiar, R.R. Yager, Melting probability measure with OWA operator to generate fuzzy measure: the crescent method. IEEE Trans. Fuzzy Syst. (2018). https://doi.org/10.1109/TFUZZ.2018.2877605
5. M. Grabisch, *Set Functions, Games and Capacities in Decision Making*, 1st edn. (Springer Publishing Company, Incorporated, New York, 2016)
6. M. Sugeno, Theory of fuzzy integrals and its applications. Ph.D. thesis, Tokyo Institute of Technology, Tokyo, Japan (1974)
7. J. Dombi, T. Jónás, The λ-additive measure in a new light: the Q_ν measure and its connections with belief, probability, plausibility, rough sets, multi-attribute utility functions and fuzzy operators. Soft Comput. (2019). https://doi.org/10.1007/s00500-019-04212-y
8. J. Dombi, T. Jónás, An elementary proof of the general Poincaré formula for λ-additive measures. Acta Cybern. **24**(2), 173–185 (2019). https://doi.org/10.14232/actacyb.24.2.2019.1
9. J. Dombi, T. Jónás, Inequalities for λ-additive measures based on the application of the general Poincaré formula for λ-additive measures. Fuzzy Sets Syst. (2019). https://doi.org/10.1016/j.fss.2019.09.007
10. J. Dombi, T. Jónás, The ν-additive measure as an alternative to the λ-additive measure, in *38th Linz Seminar on Fuzzy Set Theory*, ed. by M. Grabish, T. Kroupa, S. Saminger-Platz, T. Vetterlein (2019), pp. 26–29
11. J. Dombi, T. Jónás, The general Poincaré formula for λ-additive measures. Inf. Sci. **490**, 285–291 (2019). https://doi.org/10.1016/j.ins.2019.03.059
12. J. Dombi, T. Jónás, Lower and upper bounds for the probabilistic Poincaré formula using the general Poincaré formula for λ-additive measures. Fuzzy Sets Syst. (2020). https://doi.org/10.1016/j.fss.2020.03.020
13. C. Magadum, M. Bapat, Ranking of students for admission process by using Choquet integral. Int. J. Fuzzy Math. Arch. **15**(2), 105–113 (2018)
14. M.A. Mohamed, W. Xiao, Q-measures: an efficient extension of the Sugeno λ-measure. IEEE Trans. Fuzzy Syst. **11**(3), 419–426 (2003)
15. I. Chiţescu, Why λ-additive (fuzzy) measures? Kybernetika **51**(2), 246–254 (2015)
16. X. Chen, Y.-A. Huang, X.-S. Wang, Z.-H. You, K.C. Chan, FMLNCSIM: fuzzy measure-based lncRNA functional similarity calculation model. Oncotarget **7**(29), 45948–45958 (2016). https://doi.org/10.18632/oncotarget.10008

17. A.K. Singh, Signed λ-measures on effect algebras, in *Proceedings of the National Academy of Sciences, India Section A: Physical Sciences* (Springer India, 2018), pp. 1–7. https://doi.org/10.1007/s40010-018-0510-x
18. H. Wenxiu, L. Lushu, The g_λ-measures and conditional g_λ-measures on measurable spaces. Fuzzy Sets Syst. **46**(2), 211–219 (1992). https://doi.org/10.1016/0165-0114(92)90133-O
19. G. Banon, Distinction entre plusieurs sous-ensembles de mesures floues. Note interne **78**(11) (1978)
20. D. Dubois, H. Prade, *Fuzzy Sets and Systems: Theory and Applications*. Mathematics in Science and Engineering, vol. 144 (Academic, Orlando, 1980)
21. R.L. Keeney, Multiplicative utility functions. Oper. Res. **22**(1), 22–34 (1974)
22. J. Dombi, Towards a general class of operators for fuzzy systems. IEEE Trans. Fuzzy Syst. **16**(2), 477–484 (2008). https://doi.org/10.1109/TFUZZ.2007.905910
23. M. Sugeno, Fuzzy measures and fuzzy integrals-a survey, in *Readings in Fuzzy Sets for Intelligent Systems*, ed. by D. Dubois, H. Prade, R.R. Yager (Morgan Kaufmann, Burlington, 1993), pp. 251–257. https://doi.org/10.1016/B978-1-4832-1450-4.50027-4
24. D. Dubois, H. Prade, Rough fuzzy sets and fuzzy rough sets. Int. J. Gen. Syst. **17**(2–3), 191–209 (1990). https://doi.org/10.1080/03081079008935107
25. Y. Yao, P. Lingras, Interpretations of belief functions in the theory of rough sets. Inf. Sci. **104**(1), 81–106 (1998). https://doi.org/10.1016/S0020-0255(97)00076-5. ISSN 0020-0255
26. W.-Z. Wu, Y. Leung, W.-X. Zhang, Connections between rough set theory and Dempster-Shafer theory of evidence. Int. J. Gen. Syst. **31**(4), 405–430 (2002)
27. Z. Pawlak, Rough sets. Int. J. Comput. Inf. Sci. **11**(5), 341–356 (1982). https://doi.org/10.1007/BF01001956
28. A. Skowron, The relationship between the rough set theory and evidence theory. Bull. Pol. Acad. Sci.: Math. **37**, 87–90 (1989)
29. A. Skowron, The rough sets theory and evidence theory. Fundam. Inf. **13**(3), 245–262 (1990). ISSN 0169-2968
30. R.L. Keeney, H. Raiffa, *Decisions with Multiple Objectives: Preferences and Value Tradeoffs* (Cambridge University Press, Cambridge, 1993)
31. J. Dombi, A general class of fuzzy operators, the De Morgan class of fuzzy operators and fuzziness measures induced by fuzzy operators. Fuzzy Sets Syst. **8**(2), 149–163 (1982)
32. W. Wang, X. Liu, Intuitionistic fuzzy information aggregation using Einstein operations. IEEE Trans. Fuzzy Syst. **20**(5), 923–938 (2012)
33. H. Hamacher, *Über logische Aggregationen nicht-binär explizierter Entscheidungskriterien: Ein axiomat. Beitr. zur normativen Entscheidungstheorie* (Fischer, 1978)
34. H. Zimmermann, *Fuzzy Set Theory and Its Applications* (SpringerLink, Bücher, Springer, Netherlands, 2013)
35. L.A. Zadeh, Information and Control. Fuzzy Sets **8**(3), 338–353 (1965)

Chapter 3
The Pliant Probability Distribution Family

In this chapter, a new four-parameter probability distribution function is introduced and some of its applications are discussed. As the novel distribution function is so flexible that it may be viewed as an alternative to some notable distribution functions, we will call it the pliant distribution function. The cumulative distribution function of the novel probability distribution is founded on the so-called omega function. We will demonstrate that both the omega and the exponential function $f(x) = \mathrm{e}^{\alpha x^{\beta}}$ $(x, \alpha, \beta \in \mathbb{R}, \beta > 0)$ may be deduced from a common differential equation that we call the generalized exponential differential equation. Furthermore, we will show that the omega function, which has the α, β and d parameters $(\alpha, \beta, d \in \mathbb{R}, \beta, d > 0)$, is asymptotically identical to the exponential function $f(x) = \mathrm{e}^{\alpha x^{\beta}}$. Exploiting this result, we will elaborate on how the pliant probability distribution function can be utilized to approximate some notable probability distribution functions, the formulas of which include exponential terms. Namely, we will show that the asymptotic pliant distribution function may coincide with the Weibull-, exponential and logistic probability distribution functions. Moreover, we will demonstrate that with appropriate parameter settings, the pliant probability distribution function can approximate the standard normal probability distribution function quite well, while the approximating formula is very simple and contains only one parameter. Here, we will also highlight the interesting fact that this approximation formula may be viewed as a special case of a modifier operator in the continuous-valued logic. The flexibility of the novel probability distribution function lays the foundations for its applications in many fields of science and in a wide range of modeling problems. We will point out that the pliant probability distribution function, as an alternative to the Weibull- and exponential distribution functions, can be utilized for modeling constant, monotonic and bathtub-shaped hazard functions in reliability theory. We will also demonstrate that a function transformed from the new probability distribution function can be

J. Dombi and T. Jónás, *Advances in the Theory of Probabilistic and Fuzzy Data Scientific Methods with Applications*, Studies in Computational Intelligence 814,
https://doi.org/10.1007/978-3-030-51949-0_3

applied in the so-called kappa regression analysis method, which may be viewed as an alternative to logistic regression. This chapter is connected with the following publications of ours: [1–4].

3.1 The Pliant Probability Distribution Function

Now, we will introduce a new four-parameter probability distribution function which we call the pliant probability distribution function. This novel function has the parameters α, β, γ and d, where $\alpha > 0$, $d > 0$, $\gamma \in \{-1, 1\}$ and $\beta \in B_\gamma$. First of all, we will define the domain B_γ of parameter β.

Definition 3.1 The set B_γ is given by

$$B_\gamma = \{b^{\frac{1}{2}(\gamma+1)} : b \in \mathbb{R}^+, \gamma \in \{-1, 1\}\}. \tag{3.1}$$

Notice that if $\gamma = 1$, then $B_\gamma = \mathbb{R}^+$, and if $\gamma = -1$, then $B_\gamma = \{1\}$. From here on, a probability distribution function is always a cumulative distribution function (CDF). The pliant probability distribution function is based on an auxiliary function that we call the omega function, the appropriate linear transformation of which is the generator function of certain unary operators in continuous-valued logic [5]. Firstly, we will introduce the omega function.

Definition 3.2 The omega function $\omega_d^{(\alpha,\beta)}(x)$ is given by

$$\omega_d^{(\alpha,\beta)}(x) = \left(\frac{d^\beta + x^\beta}{d^\beta - x^\beta}\right)^{\frac{\alpha d^\beta}{2}}, \tag{3.2}$$

where $\alpha, d \in \mathbb{R}$, $d > 0$, $\beta \in B_\gamma$, $x \in \left(\frac{d}{2}(\gamma - 1), d\right)$, $\gamma \in \{-1, 1\}$.

Later we will explain why this formula is so useful. Utilizing the omega function, we define the pliant probability distribution function.

Definition 3.3 The pliant probability distribution function $F_P(x; \alpha, \beta, \gamma, d)$ is given by

$$F_P(x; \alpha, \beta, \gamma, d) = \begin{cases} 0, & \text{if } x \leq \frac{d}{2}(\gamma - 1) \\ \left(1 - \gamma\omega_d^{(-\alpha,\beta)}(x)\right)^\gamma, & \text{if } x \in \left(\frac{d}{2}(\gamma - 1), d\right) \\ 1, & \text{if } x \geq d, \end{cases} \tag{3.3}$$

where $\alpha, d \in \mathbb{R}$, $\alpha > 0$, $d > 0$, $\beta \in B_\gamma$, $\gamma \in \{-1, 1\}$.

In order to demonstrate that the function $F_P(x; \alpha, \beta, \gamma, d)$ is in fact a probability distribution function of a continuous random variable, we will discuss the main properties of the omega function.

3.1.1 *Main Properties of the Omega Function*

Here, we state the most important properties of the omega function, namely domain, differentiability, monotonity, limits and convexity [2].

Domain We will utilize the omega function either with the domain $x \in (0, d)$, or with the domain $x \in (-d, d)$. Note that the domain B_γ of parameter β is connected with the domain of x; that is, since $x \in \left(\frac{d}{2}(\gamma - 1), d\right)$, $\beta \in B_\gamma$ and $\gamma \in \{-1, 1\}$, one of the following two cases holds:

- if $\gamma = 1$, then $x \in (0, d)$ and $\beta > 0$
- if $\gamma = -1$, then $x \in (-d, d)$ and $\beta = 1$.

Notice that we allow the parameter β to have the value of 1 when $x \in (0, d)$, but β just has a value of 1 when $x \in (-d, d)$.

Differentiability $\omega_d^{(\alpha,\beta)}(x)$ is differentiable in the interval $\left(\frac{d}{2}(\gamma - 1), d\right)$.

Monotonicity

- If $\alpha > 0$, then $\omega_d^{(\alpha,\beta)}(x)$ is strictly monotonously increasing
- If $\alpha < 0$, then $\omega_d^{(\alpha,\beta)}(x)$ is strictly monotonously decreasing
- If $\alpha = 0$, then $\omega_d^{(\alpha,\beta)}(x)$ has a constant value of 1

in the interval $\left(\frac{d}{2}(\gamma - 1), d\right)$.

Limits

- If $x \in (0, d)$ and $\beta > 0$, then

$$\lim_{x \to 0^+} \omega_d^{(\alpha,\beta)}(x) = 1,$$

$$\lim_{x \to d^-} \omega_d^{(\alpha,\beta)}(x) = \begin{cases} \infty, & \text{if } \alpha > 0 \\ 0, & \text{if } \alpha < 0. \end{cases}$$

- If $x \in (-d, d)$ and $\beta = 1$, then

$$\lim_{x \to -d^+} \omega_d^{(\alpha,\beta)}(x) = \begin{cases} 0, & \text{if } \alpha > 0 \\ \infty, & \text{if } \alpha < 0, \end{cases}$$

$$\lim_{x \to d^-} \omega_d^{(\alpha,\beta)}(x) = \begin{cases} \infty, & \text{if } \alpha > 0 \\ 0, & \text{if } \alpha < 0. \end{cases}$$

Convexity It can be shown that if $x \in (0, d)$ and $\beta > 0$, then the convexity of the function $\omega_d^{(\alpha,\beta)}(x)$ in the interval $(0, d)$ is as follows:

- If

$$d^{2\beta} < \frac{4\left(\beta^2 - 1\right)}{\alpha^2 \beta^2}, \alpha \neq 0,$$

then $\omega_d^{(\alpha,\beta)}(x)$ is convex when $\alpha > 0$ and $\omega_d^{(\alpha,\beta)}(x)$ is concave when $\alpha < 0$.

- If

$$d^{2\beta} \geq \frac{4\left(\beta^2 - 1\right)}{\alpha^2 \beta^2}, \alpha \neq 0,$$

then we can distinguish the following cases:

  - if $\alpha > 0$ and $0 < \beta < 1$, then $\omega_d^{(\alpha,\beta)}(x)$ changes its shape from concave to convex at x_r
  - if $\alpha > 0$ and $\beta \geq 1$, then $\omega_d^{(\alpha,\beta)}(x)$ is convex
  - if $\alpha < 0$, $0 < \beta \leq 1$ and $x_r < d$, then $\omega_d^{(\alpha,\beta)}(x)$ changes its shape from convex to concave at x_r
  - if $\alpha < 0$, $0 < \beta \leq 1$ and $x_r \geq d$, then $\omega_d^{(\alpha,\beta)}(x)$ is convex
  - if $\alpha < 0$, $\beta > 1$ and $x_r < d$, then $\omega_d^{(\alpha,\beta)}(x)$ changes its shape from concave to convex at x_l and from convex to concave at x_r
  - if $\alpha < 0$, $\beta > 1$ and $x_r \geq d$, then $\omega_d^{(\alpha,\beta)}(x)$ changes its shape from concave to convex at x_l,

where

$$x_l = \left(\frac{-\alpha \beta d^{2\beta} - \sqrt{\alpha^2 \beta^2 d^{4\beta} - 4(\beta^2 - 1) d^{2\beta}}}{2(\beta + 1)} \right)^{1/\beta}$$

$$x_r = \left(\frac{-\alpha \beta d^{2\beta} + \sqrt{\alpha^2 \beta^2 d^{4\beta} - 4(\beta^2 - 1) d^{2\beta}}}{2(\beta + 1)} \right)^{1/\beta}.$$

It can be also shown that if $x \in (-d, d)$ and $\beta = 1$, then the shape of the function $\omega_d^{(\alpha,\beta)}(x)$ in the interval $(-d, d)$ is as follows:

- if $|\alpha| d \geq 2$, then $\omega_d^{(\alpha,\beta)}(x)$ is convex
- if $\alpha > 0$ and $\alpha d < 2$, then $\omega_d^{(\alpha,\beta)}(x)$ changes from concave to convex at $-\alpha d^2/2$
- if $\alpha < 0$ and $\alpha d > -2$, then $\omega_d^{(\alpha,\beta)}(x)$ changes from convex to concave at $-\alpha d^2/2$.

Based on the above-mentioned properties of the omega function, the basic semantics of the parameters α, β and d can be summarized as follows. The parameter d determines the domain of the omega function (either $x \in (-d, d)$, or $x \in (0, d)$), the parameter α influences its monotonicity and steepness, while the parameter β also affects the steepness of $\omega_d^{(\alpha,\beta)}(x)$, $(\alpha, d \in \mathbb{R}, d > 0, \beta > 0)$. Figure 3.1 shows plots of some omega functions.

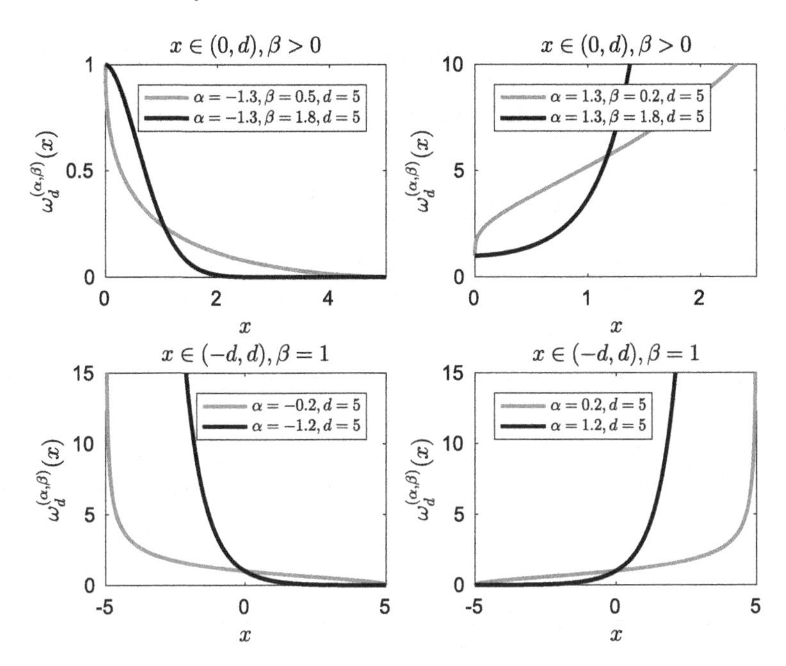

Fig. 3.1 Plots of some omega functions

3.1.1.1 The Generalized Exponential Differential Equation

Now we will introduce the generalized exponential differential equation and show how it is related to the exponential function $f(x) = \mathrm{e}^{\alpha x^{\beta}}$, ($\alpha, \beta \in \mathbb{R}, \beta > 0$) and to the omega function [2].

Definition 3.4 We define the generalized exponential differential equation as

$$\frac{\mathrm{d}f(x)}{\mathrm{d}x} = \alpha\beta x^{\beta-1}\left(\frac{d^{2\beta}}{d^{2\beta} - x^{2\beta}}\right)^{\varepsilon} f(x), \tag{3.4}$$

where $\alpha, d \in \mathbb{R}, d > 0, \beta \in B_{\gamma}, x \in \left(\frac{d}{2}(\gamma - 1), d\right), \gamma \in \{-1, 1\}, \varepsilon \in \{0, 1\}, f(x) > 0$.

Lemma 3.1 *The solutions of the generalized exponential differential equation are:*

$$f(x) = \begin{cases} C\mathrm{e}^{\alpha x^{\beta}}, & \text{if } \varepsilon = 0 \\ C\left(\frac{d^{\beta}+x^{\beta}}{d^{\beta}-x^{\beta}}\right)^{\frac{\alpha d^{\beta}}{2}}, & \text{if } \varepsilon = 1, \end{cases}$$

where $C \in \mathbb{R}, C > 0$.

Proof Recall that either $x \in (0, d)$ and $\beta > 0$ (the case of $\gamma = 1$), or $x \in (-d, d)$ and $\beta = 1$ (the case of $\gamma = -1$) holds.

If $\varepsilon = 0$, then the differential equation in Eq. (3.4) may be written as

$$\frac{\mathrm{d}f(x)}{\mathrm{d}x} = \alpha\beta x^{\beta-1} f(x), \tag{3.5}$$

Separating the variables in Eq. (3.5) and integrating both sides leads to

$$\int \frac{1}{f(x)} \mathrm{d}f(x) = \int \alpha\beta x^{\beta-1} \mathrm{d}x$$

$$\ln |f(x)| = \alpha x^{\beta} + \ln C,$$

where $C > 0$. Utilizing the fact that $f(x) > 0$, the last equation may be written as

$$f(x) = C\mathrm{e}^{\alpha x^{\beta}}.$$

If $\varepsilon = 1$, then the differential equation in Eq. (3.4) becomes

$$\frac{\mathrm{d}f(x)}{\mathrm{d}x} = \alpha\beta x^{\beta-1} \frac{d^{2\beta}}{d^{2\beta} - x^{2\beta}} f(x). \tag{3.6}$$

Next, separating the variables in Eq. (3.6) and exploiting the fact that

$$\frac{1}{d^{2\beta} - x^{2\beta}} = \frac{1}{2d^{\beta}} \left(\frac{1}{d^{\beta} + x^{\beta}} + \frac{1}{d^{\beta} - x^{\beta}} \right),$$

we get

$$\int \frac{1}{f(x)} \mathrm{d}f(x) = \alpha\beta \frac{d^{\beta}}{2} \left(\int \frac{x^{\beta-1}}{d^{\beta} + x^{\beta}} \mathrm{d}x + \int \frac{x^{\beta-1}}{d^{\beta} - x^{\beta}} \mathrm{d}x \right).$$

Integrating both sides results in the equation

$$\ln |f(x)| = \frac{\alpha d^{\beta}}{2} \left(\ln |d^{\beta} + x^{\beta}| - \ln |d^{\beta} - x^{\beta}| \right) + \ln C,$$

where $C > 0$. Since $f(x) > 0$ and

- $x \in (-d, d)$, $\beta = 1$, if $\gamma = -1$
- $x \in (0, d)$, $\beta > 0$, if $\gamma = 1$.

The last equation may be written as

$$f(x) = C \left(\frac{d^{\beta} + x^{\beta}}{d^{\beta} - x^{\beta}} \right)^{\frac{\alpha d^{\beta}}{2}}.$$

□

3.1.1.2 Connections Between the Exponential and Omega Functions

Lemma 3.1 suggests that there is a key connection between the exponential function $f(x) = \mathrm{e}^{\alpha x^\beta}$ and the omega function. Namely, the solution of the generalized exponential differential equation with $\varepsilon = 0$ and $C = 1$ is simply the exponential function $f(x) = \mathrm{e}^{\alpha x^\beta}$, while the solution of Eq. (3.4) with $\varepsilon = 1, C = 1$ is the omega function. Furthermore, if d is much greater than x, then

$$\frac{d^{2\beta}}{d^{2\beta} - x^{2\beta}} \approx 1.$$

In this case, the generalized exponential differential equation for $\varepsilon = 1$ becomes the following approximate equation:

$$\frac{\mathrm{d}f(x)}{\mathrm{d}x} \approx \alpha\beta x^{\beta-1} f(x),$$

which is nearly the generalized exponential differential equation with $\varepsilon = 0$, the solution of which is the exponential function $f(x) = \mathrm{e}^{\alpha x^\beta}$. The following theorem provides the theoretical basis for this result [2].

Theorem 3.1 *For any* $x \in \left(\frac{d}{2}(\gamma - 1), d\right)$,

$$\lim_{d\to\infty} \omega_d^{(\alpha,\beta)}(x) = \mathrm{e}^{\alpha x^\beta}, \tag{3.7}$$

where $\alpha, d \in \mathbb{R}$, $d > 0$, $\beta \in B_\gamma$, $\gamma \in \{-1, 1\}$.

Proof Here again either $x \in (0, d)$ and $\beta > 0$ (the case of $\gamma = 1$), or $x \in (-d, d)$ and $\beta = 1$ (the case of $\gamma = -1$) holds.

Let x have a fixed value in the interval $\left(\frac{d}{2}(\gamma - 1), d\right)$.

$$\lim_{d\to\infty} \omega_d^{(\alpha,\beta)}(x) = \lim_{d\to\infty} \left(\frac{d^\beta + x^\beta}{d^\beta - x^\beta}\right)^{\frac{\alpha d^\beta}{2}} = \lim_{d\to\infty} \left(\left(\frac{d^\beta - x^\beta + 2x^\beta}{d^\beta - x^\beta}\right)^{d^\beta}\right)^{\frac{\alpha}{2}} =$$

$$= \lim_{d\to\infty} \left(\left(1 + \frac{2x^\beta}{d^\beta - x^\beta}\right)^{d^\beta}\right)^{\frac{\alpha}{2}}.$$

Since x is fixed, if $d \to \infty$, then $\Delta = d^\beta - x^\beta \to \infty$ and so the previous calculation can be continued as follows.

$$\lim_{d\to\infty}\left(\left(1+\frac{2x^{\beta}}{d^{\beta}-x^{\beta}}\right)^{d^{\beta}}\right)^{\frac{\alpha}{2}} = \lim_{\Delta\to\infty}\left(\left(1+\frac{2x^{\beta}}{\Delta}\right)^{\Delta+x^{\beta}}\right)^{\frac{\alpha}{2}} =$$

$$= \left(\lim_{\Delta\to\infty}\left(1+\frac{2x^{\beta}}{\Delta}\right)^{\Delta}\lim_{\Delta\to\infty}\left(1+\frac{2x^{\beta}}{\Delta}\right)^{x^{\beta}}\right)^{\frac{\alpha}{2}} =$$

$$= \left(e^{2x^{\beta}}\right)^{\frac{\alpha}{2}}\cdot 1^{\frac{\alpha}{2}} = e^{\alpha x^{\beta}}.$$

□

Based on Theorem 3.1, we can state that the asymptotic omega function is just the exponential function $f(x) = e^{\alpha x^{\beta}}$. Actually, if $x \ll d$, then $\omega_d^{(\alpha,\beta)}(x) \approx e^{\alpha x^{\beta}}$; that is, if d is sufficiently large, then the omega function suitably approximates the exponential function $f(x) = e^{\alpha x^{\beta}}$.

3.2 Approximations and Their Applications

Here, we will demonstrate how the pliant probability distribution function $F_P(x; \alpha, \beta, \gamma, d)$ can be utilized for approximating the Weibull-, exponential, logistic and standard normal probability distribution functions. First, we will show that the function $F_P(x; \alpha, \beta, \gamma, d)$ is in fact a probability distribution function.

Lemma 3.2 *The function $F_P(x; \alpha, \beta, \gamma, d)$ given by Definition 3.3 is a probability distribution function.*

Proof The function $F_P(x; \alpha, \beta, \gamma, d)$ is a probability distribution function if it is:

(1) monotonously increasing
(2) left continuous
(3) and $\lim_{x\to-\infty} F_P(x; \alpha, \beta, \gamma, d) = 0$ and $\lim_{x\to+\infty} F_P(x; \alpha, \beta, \gamma, d) = 1$.

These properties of the function $F_P(x; \alpha, \beta, \gamma, d)$ trivially follow from the properties of the omega function $\omega_d^{(\alpha,\beta)}(x)$. □

We have demonstrated that the omega function in the asymptotic limit is just the exponential function $f(x) = e^{\alpha x^{\beta}}$, where $\beta > 0$. Exploiting this result and the fact the function $F_P(x; \alpha, \beta, \gamma, d)$ is a probability distribution function, we will show how the pliant probability distribution function can be utilized for approximating some notable probability distribution functions that include exponential terms. From now on, we will use the notation $\xi \sim D_p(\alpha, \beta, \gamma, d)$ to indicate that the random variable ξ has a pliant probability distribution with the parameters α, β, γ, d. That is, if $\xi \sim D_p(\alpha, \beta, \gamma, d)$, then $P(\xi < x) = F_P(x; \alpha, \beta, \gamma, d)$ for any $x \in \mathbb{R}$.

3.2.1 Approximation to the Weibull Probability Distribution Function

The 2-parameter distribution function $F_W(x; \beta, \lambda)$ of the random variable which has a Weibull probability distribution is commonly given by

$$F_W(x; \beta, \lambda) = \begin{cases} 0, & \text{if } x \leq 0 \\ 1 - \mathrm{e}^{-(x/\lambda)^\beta} & \text{if } x > 0, \end{cases} \tag{3.8}$$

where $\beta, \lambda \in \mathbb{R}$ and $\beta, \lambda > 0$ are the shape and scale parameters of the distribution, respectively [6, 7]. By using the substitution $\alpha = \lambda^{-\beta}$, Eq. (3.8) may be written in the form

$$F_W(x; \alpha, \beta) = \begin{cases} 0, & \text{if } x \leq 0 \\ 1 - \mathrm{e}^{-\alpha x^\beta}, & \text{if } x > 0, \end{cases} \tag{3.9}$$

where $\alpha, \beta \in \mathbb{R}$, $\alpha, \beta > 0$. Hereafter, we will use this alternative definition of the 2-parameter distribution function of the random variable that has a Weibull probability distribution. Furthermore, we will use the notation $\eta \sim W(\alpha, \beta)$ to indicate that the random variable η has a Weibull probability distribution with the parameters $\alpha, \beta > 0$; that is, if $\eta \sim W(\alpha, \beta)$, then $P(\eta < x) = F_W(x; \alpha, \beta)$ for any $x \in \mathbb{R}$.

Theorem 3.2 *If $\xi \sim D_p(\alpha, \beta, \gamma, d)$, $\eta \sim W(\alpha, \beta)$ and $\gamma = 1$, then for any $x \in \mathbb{R}$,*

$$\lim_{d \to \infty} P(\xi < x) = P(\eta < x),$$

where $\alpha, d \in \mathbb{R}$, $\alpha, d > 0$, $\beta \in B_\gamma$ [2].

Proof Utilizing the definitions of $F_P(x; \alpha, \beta, \gamma, d)$ and B_γ, if $\gamma = 1$, then the pliant probability distribution function $F_P(x; \alpha, \beta, \gamma, d)$ may be written as

$$F_P(x; \alpha, \beta, \gamma, d) = \begin{cases} 0, & \text{if } x \leq 0 \\ 1 - \omega_d^{(-\alpha,\beta)}(x), & \text{if } x \in (0, d) \\ 1, & \text{if } x \geq d. \end{cases} \tag{3.10}$$

Let $x \in \mathbb{R}$ be fixed. We will now distinguish the following two cases.

(1) If $x \leq 0$, then $F_P(x; \alpha, \beta, \gamma, d) = F_W(x; \beta, \lambda)$ holds by definition.
(2) If $x \in (0, d)$, $d > 0$, then

$$F_P(x; \alpha, \beta, \gamma, d) = 1 - \omega_d^{(-\alpha,\beta)}(x).$$

Next, following Theorem 3.1, if $d \to \infty$, then

$$F_P(x; \alpha, \beta, \gamma, d) = 1 - \omega_d^{(-\alpha,\beta)}(x) \to 1 - \mathrm{e}^{-\alpha x^{\beta}} = F_W(x; \alpha, \beta).$$

That is,

$$\lim_{d \to \infty} P(\xi < x) = P(\eta < x).$$

□

Note that we call the probability distribution given by the pliant CDF in Eq. (3.10) the omega probability distribution [2]. Following our results in [2], Okorie and Nadarajah showed the main properties of the omega probability distribution in [8].

3.2.1.1 Statistical Estimation of Parameters

Here, we will discuss two methods for the parameter estimation of the omega distribution. Firstly, we will provide the log-likelihood function and propose a method to maximize it. Secondly, we will discuss how the parameters of the omega distribution can be estimated by fitting its CDF to an empirical CDF. Here, we assume that the random variable τ represents the time-to-first-failure of a component or system and τ has an omega probability distribution with the parameters $\alpha, \beta, d > 0$. It should be added that the above-mentioned two methods utilize different data set types. In the case of maximum likelihood estimation, we assume that independent and identically distributed $t_1, t_2, \ldots, t_n$ observations are available on the random variable τ. However, in many cases of practical reliability engineering, the exact $t_1, t_2, \ldots, t_n$ time-to-first failure data are not available; instead we have frequency data indicating the number of components or systems that have failed in given time periods. In such cases, the empirical CDF of τ can be directly produced and the second method, which estimates the parameters by fitting the omega CDF to the empirical CDF of τ, can be applied. We will use the notation $\tau \sim \omega(\alpha, \beta, d)$ to indicate that the random variable τ has an omega probability distribution with the parameters $\alpha, \beta, d > 0$.

Maximum Likelihood Estimation

Let $t_1, t_2, \ldots, t_n$ be independent and identically distributed observations on the random variable τ and $\tau \sim \omega(\alpha, \beta, d)$. The omega probability density function $f_d^{(\alpha,\beta)}(x)$ of τ is

$$f_d^{(\alpha,\beta)}(x) = \begin{cases} 0, & \text{if } x \leq 0 \\ \alpha\beta x^{\beta-1} \frac{d^{2\beta}}{d^{2\beta} - x^{2\beta}} \omega_d^{(-\alpha,\beta)}(x), & \text{if } 0 < x < d \\ 0, & \text{if } x \geq d, \end{cases}$$

where $\alpha, \beta, d > 0$. The likelihood function $L(\alpha, \beta, d)$ is

$$L(\alpha,\beta,d)=\prod_{i=1}^{n} f_d^{(\alpha,\beta)}(t_i)=$$

$$=\alpha^n\beta^n\prod_{i=1}^{n}\left(t_i^{\beta-1}\frac{d^{2\beta}}{d^{2\beta}-t_i^{2\beta}}\left(\frac{d^\beta+t_i^\beta}{d^\beta-t_i^\beta}\right)^{\frac{-\alpha d^\beta}{2}}\right).$$

The log-likelihood function $l(\alpha,\beta,d)=\ln(L(\alpha,\beta,d))$ is

$$\begin{aligned}l(\alpha,\beta,d)&=n\ln\alpha+n\ln\beta+(\beta-1)\sum_{i=1}^{n}\ln t_i+\\&+\sum_{i=1}^{n}\ln\frac{d^{2\beta}}{d^{2\beta}-t_i^{2\beta}}-\frac{\alpha d^\beta}{2}\sum_{i=1}^{n}\ln\frac{d^\beta+t_i^\beta}{d^\beta-t_i^\beta}.\end{aligned}\tag{3.11}$$

Notice that the parameter d specifies the support of $f_d^{(\alpha,\beta)}(x)$; that is, $f_d^{(\alpha,\beta)}(x)$ is positive only if $x\in(0,d)$. It means that the value of parameter d needs to satisfy the condition $d>\max_{i=1,\ldots,n}(t_i)$. So the maximum likelihood estimations of the parameters α, β and d can be obtained by solving the following minimization problem:

$$\begin{gathered}-l(\alpha,\beta,d)\to\min\\\alpha,\beta,d>0\\d>\max_{i=1,\ldots,n}(t_i).\end{gathered}$$

There is no closed form solution for this minimization problem, but it can be solved by utilizing a global optimization method. We propose the application of the so-called GLOBAL method, which is a stochastic global optimization procedure introduced by Csendes et al. [9, 10].

It is worth mentioning that there is an interesting connection between the log-likelihood functions of the Weibull and omega probability distributions. On the one hand, the log-likelihood function of the Weibull probability distribution given by the CDF in Eq. (3.9) is

$$l(\alpha,\beta)=n\ln\alpha+n\ln\beta+(\beta-1)\sum_{i=1}^{n}\ln t_i-\alpha\sum_{i=1}^{n}t_i^\beta.\tag{3.12}$$

On the other hand,

$$\lim_{d\to\infty}\left(\sum_{i=1}^{n}\ln\frac{d^{2\beta}}{d^{2\beta}-t_i^{2\beta}}\right)=0,$$

and based on Theorem 3.1,

$$\lim_{d\to\infty}\left(\frac{\alpha d^{\beta}}{2}\sum_{i=1}^{n}\ln\frac{d^{\beta}+t_i^{\beta}}{d^{\beta}-t_i^{\beta}}\right)=\alpha\sum_{i=1}^{n}t_i^{\beta}.$$

Hence, if $d\to\infty$, then the log-likelihood function of the omega probability distribution in Eq. (3.11) is identical to the log-likelihood function of the Weibull probability distribution given by Eq. (3.12).

Fitting the Cumulative Probability Distribution Function

Let $N(t)$ denote the number of components or systems that have survived up to time t from the number of components or systems $N(0)$ that were initially put into operation. Let Δt denote the length of a time period, and let $t=i\Delta t$, where $i=0,1,\ldots,n$ and $\Delta t>0$. If $\Delta t=1$, then the $N(t)-N(t+\Delta t)$ difference, which represents the number of components or systems that fail in the time interval $(t,t+\Delta t]$, is $N(i)-N(i+1)$, where $i=0,1,\ldots,n-1$. For example, if $\Delta t=1$ week, then the difference $N(3)-N(4)$ represents the number of components or systems that failed on the 4th week. As noted before, there are cases in practice where the exact $t_1,t_2,\ldots,t_n$ time-to-first-failure data are not available; rather frequency data values are available indicating the number of components or systems that have failed in given time periods. In such cases, when the $N(0),N(1),\ldots,N(n)$ data values are available, the empirical cumulative distribution function $\hat{F}(t)$ of the time-to-first-failure random variable τ can be computed as

$$\hat{F}(i)=1-\frac{N(i)}{N(0)},\tag{3.13}$$

where $i=0,1,\ldots,n$. Next, the parameters α, β and d of the omega probability distribution can be identified by fitting the omega CDF $F_d^{(\alpha,\beta)}(t)$ to the empirical CDF $\hat{F}(t)$. For this purpose we need to minimize the following sum of squares:

$$S(\alpha,\beta,d)=\sum_{i=1}^{n}\left(F_d^{(\alpha,\beta)}(i)-\hat{F}(i)\right)^2,\tag{3.14}$$

with the constraints $\alpha,\beta,d>0$. This minimization problem can be solved by utilizing the GLOBAL method that we referenced earlier in this section. In our demonstrative example, we will show how the CDF fitting method can be applied in practice.

3.2.1.2 Applications in Reliability Theory

Let the continuous random variable τ be the time-to-first-failure of a component or system. In reliability theory, the failure rate function $h(t)$ for τ is given by

$$h(t)=\lim_{\Delta t\to 0}\frac{F(t+\Delta t)-F(t)}{\Delta t R(t)}=\frac{f(t)}{R(t)},\tag{3.15}$$

where $f(t)$ is the probability density function of τ. The hazard function $h(t)$ is also called the failure rate function. In practice, the quantity $h(t)\Delta t$ represents the conditional probability that a component or a system will fail in the time interval $(t, t+\Delta t]$, given that it has survived up to time t, $(t,\ \Delta t > 0)$.

A typical hazard function curve of a component or a system is bathtub-shaped; that is, it can be divided into three distinct phases called the infant mortality period, useful life, and wear-out period. It is typical that the probability distribution of τ is different in the three characteristic phases of the bathtub-shaped hazard function. If τ has a Weibull probability distribution with the parameters $\alpha, \beta > 0$, then using Eq. (3.15), the hazard function $h_W(t; \alpha, \beta)$ of τ (which we call the Weibull hazard function) is

$$h_W(t; \alpha, \beta) = \frac{f_W(t; \alpha, \beta)}{1 - F_W(t; \alpha, \beta)} = \frac{\alpha\beta t^{\beta-1}\mathrm{e}^{-\alpha t^{\beta}}}{\mathrm{e}^{-\alpha t^{\beta}}} = \alpha\beta t^{\beta-1}, \tag{3.16}$$

where $f_W(t; \alpha, \beta)$ is the probability density function of τ. Equation (3.16) suggests an important property of the hazard function $h(t; \alpha, \beta)$. Namely,

- if $0 < \beta < 0$, then $h_W(t; \alpha, \beta)$ is decreasing
- if $\beta = 1$, then $h_W(t; \alpha, \beta)$ is constant with the value of α
- if $\beta > 1$, then $h_W(t; \alpha, \beta)$ is increasing

with respect to time. That is, if $\tau \sim W(\alpha, \beta)$, then the failure rate function $h_W(t; \alpha, \beta)$, with appropriate values of its parameters, can characterize each of the three phases of a bathtub-shaped failure rate curve quite well. On the one hand, this flexibility of the Weibull probability distribution makes it suitable for modeling the probability distribution of time-to-first-failure random variable in a wide range of reliability analyses [11]. On the other hand, the Weibull hazard function $h_W(t; \alpha, \beta)$ is either monotonic or constant; that is, its curve cannot be bathtub-shaped. However, lifetime data of a component or system typically require non-monotonic shapes like the bathtub shape. There have been many modifications developed for the Weibull probability distribution in order to obtain non-monotonic shapes. See, for example, the publications [12–23]. A comprehensive review of the known modifications to the Weibull probability distribution can be found in a study [24]. Here, we will demonstrate that the hazard function of the pliant probability distribution can be applied to model both monotonic and bathtub-shaped hazard rate curves.

Now let us assume that τ has a pliant probability distribution with the parameters $\alpha, d > 0, \gamma = 1$. In this case, $\beta > 0$ and the hazard function of τ, which we will call the pliant hazard function, is

$$\begin{aligned} h_P(t; \alpha, \beta, d) &= \frac{f_P(t; \alpha, \beta, d)}{1 - F_P(t; \alpha, \beta, d)} = \frac{\alpha\beta t^{\beta-1}\frac{d^{2\beta}}{d^{2\beta}-t^{2\beta}}\omega_d^{(-\alpha,\beta)}(t)}{\omega_d^{(-\alpha,\beta)}(t)} = \\ &= \alpha\beta t^{\beta-1}\frac{d^{2\beta}}{d^{2\beta}-t^{2\beta}}, \end{aligned} \tag{3.17}$$

if $0 < t < d$, where $f_P(t; \alpha, \beta, d)$ is the probability density function of τ. Utilizing Eqs. (3.16) and (3.17), the pliant hazard function $h_P(t; \alpha, \beta, d)$ may be written as

$$h_P(t; \alpha, \beta, d) = h_W(\alpha, \beta) g(t; \beta, d), \tag{3.18}$$

where

$$g(t; \beta, d) = \frac{d^{2\beta}}{d^{2\beta} - t^{2\beta}},$$

and $\alpha, \beta, d > 0$, $t \in (0, d)$. That is, the pliant hazard function may be viewed as the Weibull hazard function multiplied by the corrector function $g(t; \beta, d)$. The pliant hazard function $h_P(t; \alpha, \beta, d)$, which we also call the omega hazard function, has some key properties that make it suitable for modeling bathtub-shaped failure rate curves. The following lemma allows us to utilize the pliant hazard function as an alternative to the Weibull hazard function.

Lemma 3.3 *For any $t \in (0, d)$, if $d \to \infty$, then $h_P(t; \alpha, \beta, d) \to h_W(t; \alpha, \beta)$, where $\alpha, \beta, d > 0$.*

Proof If $t \in (0, d)$ is fixed and $d \to \infty$, then $g(t; \beta, d) \to 1$ and so

$$h_P(t; \alpha, \beta, d) = h_W(\alpha, \beta) g(t; \beta, d) \to h_W(\alpha, \beta).$$

□

The practical implication of this result is as follows. Since the Weibull hazard function can be utilized as a model for each phase of a bathtub-shaped failure rate curve and $h_P(t; \alpha, \beta, d) \approx h_W(\alpha, \beta)$, if t is small compared to d, the pliant hazard function can also model each phase of the same bathtub-shaped failure rate curve, if d is sufficiently large. That is, the pliant hazard function, as an alternative to the Weibull hazard function, can be utilized as a phase-by-phase model of a bathtub-shaped failure rate curve.

Figure 3.2 shows how the Weibull- and pliant hazard function curves can model each characteristic phase of a failure rate curve. The plots in Fig. 3.2 demonstrate the results of the previous lemma; that is, if $t \ll d$, then the pliant hazard function approximates quite well the Weibull hazard function.

The next lemma demonstrates that the pliant hazard function $h_P(t; \alpha, \beta, d)$ can be utilized as a model for all the three phases of a bathtub-shaped failure rate curve.

Lemma 3.4 *If $0 < \beta < 1$, then $h_P(t; \alpha, \beta, d)$ is strictly convex in the interval $(0, d)$ and $h_P(t; \alpha, \beta, d)$ has its minimum at*

$$t_0 = d \left(\frac{1 - \beta}{1 + \beta} \right)^{\frac{1}{2\beta}}.$$

Proof The lemma follows from the elementary properties of the pliant hazard function $h_P(t; \alpha, \beta, d)$ by using its first and second derivatives. □

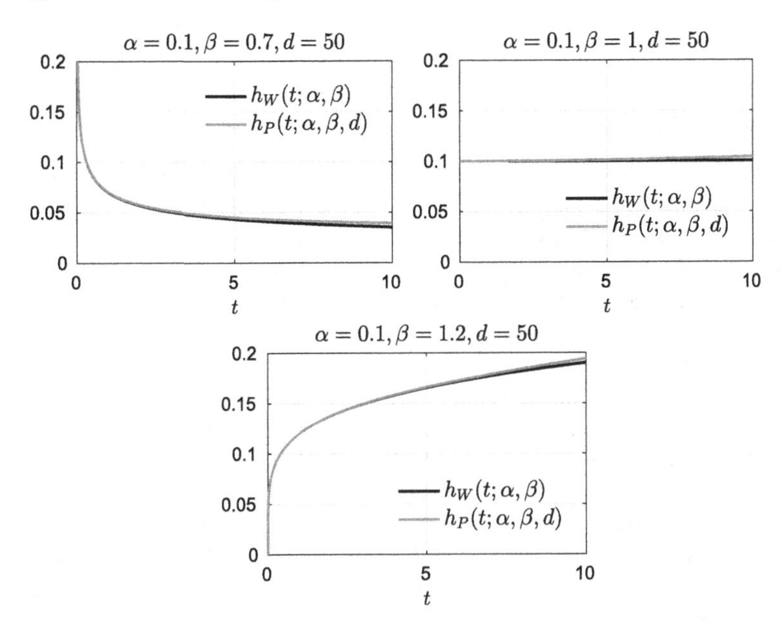

Fig. 3.2 Plots of Weibull- and pliant hazard functions

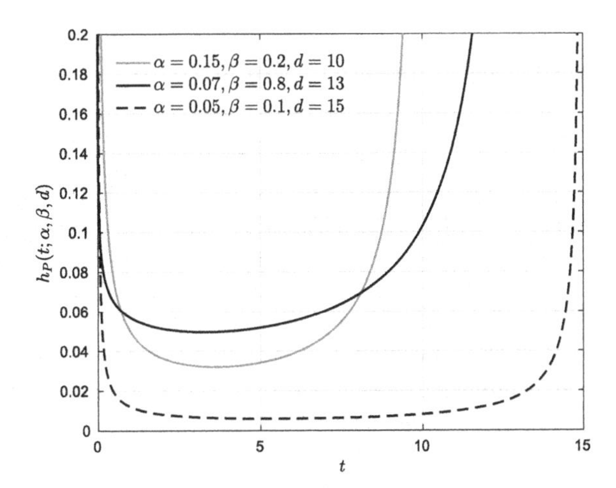

Fig. 3.3 Examples of bathtub-shaped pliant hazard function plots

Figure 3.3 shows some plots of the pliant hazard function $h_P(t; \alpha, \beta, d)$ with $0 < \beta < 1$.

Based on the above properties of the pliant hazard function, we may conclude that the pliant probability distribution with $\gamma = 1$ can be employed to describe the probability distribution of the time-to-first-failure random variable in each characteristic phase of a bathtub-shaped failure rate curve. Moreover, we have two possibilities for modeling a bathtub-shaped failure rate curve. Namely, either we piecewise describe each phase by a pliant hazard function, or we apply one pliant hazard function that models the entire failure rate curve [2].

A Demonstrative Example[1]

Here $F_d^{(\alpha,\beta)}(t)$ and $h_d^{(\alpha,\beta)}(t)$ will denote the CDF and the hazard function of the omega probability distribution; that is, $F_d^{(\alpha,\beta)}(t) = F_P(t; \alpha, \beta, \gamma, d)$ and $h_d^{(\alpha,\beta)}(t) = h_P(t; \alpha, \beta, d)$, where $\alpha, \beta, d > 0, \gamma = 1$.

Now we will demonstrate how the omega probability distribution can be utilized to model the probability distribution of the time-to-first-failure random variable if its hazard function is bathtub-shaped. In our example, real-life empirical failure data values of a laptop motherboard type were analyzed. Here, the examined motherboard type was taken into account as a component, the typical lifetime of which is between 4 and 6 years. In our case, the initially released number of motherboards was $N(0) = 21000$, while the $N(t)$ values were available on weekly basis; that is, $N(i)$ represents the number of components that have survived up to the end of the ith week, $i = 0, 1, \ldots, 295$. Using the $N(i)$ values, the empirical CDF of the time-to-first-failure random variable was calculated according to Eq. (3.13). Applying the life table method (see, e.g. Saunders [25]), the empirical hazard function (HF) $\hat{h}(t)$ can be computed as

$$\hat{h}(t) = \frac{N(t) - N(t + \Delta t)}{N(t)\Delta t}, \tag{3.19}$$

and utilizing the fact that $t = i\Delta t$, where $\Delta t = 1$, we get

$$\hat{h}(i) = \frac{N(i) - N(i + 1)}{N(i)}, \tag{3.20}$$

where $i = 0, 1, \ldots, 294$. The $N(i)$ values and the computed values of the empirical CDF and empirical HF are listed in the Tables 3.9, 3.10 and 3.11 in Appendix 3.4. Notice that from week $i = 295$ the number of functioning components $N(i)$ is zero and so $\hat{h}(i)$ could be computed for $i = 0, 1, \ldots, 294$.

Here, the parameters α, β and d of the omega probability distribution were identified by fitting the omega CDF $F_d^{(\alpha,\beta)}(t)$ to the empirical CDF $\hat{F}(t)$. For this purpose we solved the minimization problem

$$S(\alpha, \beta, d) = \sum_{i=1}^{n} \left(F_d^{(\alpha,\beta)}(i) - \hat{F}(i) \right)^2 \to \min, \tag{3.21}$$

by utilizing the GLOBAL method referenced in Sect. 3.2.1.1 (in our case $n = 295$). The GLOBAL method was implemented and the following analyses were done in the MATLAB 2017b numerical computing environment. In order to determine parameter

[1] Reprinted from József Dombi Tamás Jónás, Zsuzsanna E. Tóth, Gábor Árva, The omega probability distribution and its applications in reliability theory, Quality & Reliability Engineering International, 35(2), 600–626, https://doi.org/10.1002/qre.2425, Copyright (2019), with permission from John Wiley and Sons.

constraints for the minimization problem in Eq. (3.14) the following properties of the empirical data and of the omega CDF were taken into consideration.

The empirical hazard function $\hat{h}(t)$ in question is bathtub-shaped (see Fig. 3.5), and based on our previous results in this section, the omega hazard function $h_d^{(\alpha,\beta)}(t)$ is bathtub-shaped only if $0 < \beta < 1$. That is, we can set the constraint $0 < \beta < 1$ to find the minimum of the function $S(\alpha, \beta, d)$.

Since $F_d^{(\alpha,\beta)}(t) = 1$ if $t \geq d$, and the smallest i for which $\hat{F}(i) = 1$ holds is $i = n$, we may expect that the optimal value of parameter d is not much greater than n, and it is not much less than n. Therefore, it is reasonable to apply the constraint $n/2 \leq d \leq 2n$ for the parameter d in the minimization problem given by Eq. (3.14). Note that although this constraint setting is somewhat arbitrary, it proves to be valid in practice.

Here, the following heuristic was utilized to identify boundaries for the parameter α. Based on Lemma 3.4, if $\beta \in (0, 1)$, then the omega hazard function $h_d^{(\alpha,\beta)}(t)$ is minimal in the interval $(0, d)$ at

$$t_0 = d\left(\frac{1-\beta}{1+\beta}\right)^{\frac{1}{2\beta}},$$

and so the minimum value of $h_d^{(\alpha,\beta)}(t)$ in $(0, d)$ is

$$\min_{t\in(0,d)}\left(h_d^{(\alpha,\beta)}(t)\right) = h_d^{(\alpha,\beta)}(t_0) = \alpha\beta t_0^{\beta-1}\frac{d^{2\beta}}{d^{2\beta}-t_0^{2\beta}} =$$
$$= \alpha\frac{d^{\beta-1}}{2}(1-\beta)^{\frac{\beta-1}{2\beta}}(1+\beta)^{\frac{\beta+1}{2\beta}}.$$

By utilizing the values of the empirical hazard function $\hat{h}(t)$, we can empirically identify an h_l lower boundary and an h_u upper boundary for the minimum value of the hazard function $h_d^{(\alpha,\beta)}(t)$; that is,

$$h_l \leq \alpha\frac{d^{\beta-1}}{2}(1-\beta)^{\frac{\beta-1}{2\beta}}(1+\beta)^{\frac{\beta+1}{2\beta}} \leq h_u. \tag{3.22}$$

We will utilize the results of the following lemma to identify boundaries for the parameter α.

Lemma 3.5 *For any* $\beta \in (0, 1)$ *and* $d > 1$

$$\frac{1}{d} < d^{\beta-1} < 1 \tag{3.23}$$

$$1 < (1-\beta)^{\frac{\beta-1}{2\beta}} < \sqrt{e} \tag{3.24}$$

$$\sqrt{e} < (1+\beta)^{\frac{\beta+1}{2\beta}} < 2. \tag{3.25}$$

Proof The inequalities in Eq. (3.23) trivially follow from the conditions $\beta \in (0, 1)$ and $d > 1$.

Since

$$(1-\beta)^{\frac{\beta-1}{2\beta}}$$

is continuous and strictly monotonously decreasing in $(0, 1)$, to prove the inequalities in Eq. (3.24), it is sufficient to show that

$$\lim_{\beta \to 0^+} (1-\beta)^{\frac{\beta-1}{2\beta}} = \sqrt{e}$$

and

$$\lim_{\beta \to 1^-} (1-\beta)^{\frac{\beta-1}{2\beta}} = 1.$$

Let $\beta' = 1/\beta$. Then

$$\lim_{\beta \to 0^+} (1-\beta)^{\frac{\beta-1}{2\beta}} = \left(\lim_{\beta' \to \infty} \left(1 - \frac{1}{\beta'}\right)^{1-\beta'} \right)^{\frac{1}{2}} =$$

$$= \left(\lim_{\beta' \to \infty} \left(1 - \frac{1}{\beta'}\right) \right)^{\frac{1}{2}} \left(\lim_{\beta' \to \infty} \left(1 - \frac{1}{\beta'}\right)^{\beta'} \right)^{-\frac{1}{2}} = 1 \cdot \sqrt{e}.$$

Now, let $\beta' = 1 - \beta$. Utilizing this substitution and the L'Hospital rule gives

$$\lim_{\beta \to 1^-} (1-\beta)^{\frac{\beta-1}{2\beta}} = \left(\lim_{\beta' \to 0^+} e^{-\frac{\beta'}{1-\beta'} \ln \beta'} \right)^{\frac{1}{2}} = \left(e^0\right)^{\frac{1}{2}} = 1.$$

As

$$(1+\beta)^{\frac{\beta+1}{2\beta}}$$

is continuous and strictly monotonously increasing in $(0, 1)$, to prove the inequalities in Eq. (3.25), it is sufficient to show that

$$\lim_{\beta \to 0^+} (1+\beta)^{\frac{\beta+1}{2\beta}} = \sqrt{e}$$

and

$$\lim_{\beta \to 1^-} (1+\beta)^{\frac{\beta+1}{2\beta}} = 2. \tag{3.26}$$

Let $\beta' = 1/\beta$. Then

$$\lim_{\beta \to 0^+} (1+\beta)^{\frac{\beta+1}{2\beta}} = \left(\lim_{\beta' \to \infty} \left(1 + \frac{1}{\beta'}\right)^{(1+\beta')} \right)^{\frac{1}{2}} =$$

$$= \left(\lim_{\beta' \to \infty} \left(1 + \frac{1}{\beta'}\right)\right)^{\frac{1}{2}} \left(\lim_{\beta' \to \infty} \left(1 + \frac{1}{\beta'}\right)^{\beta'} \right)^{\frac{1}{2}} = \sqrt{e}.$$

Since Eq. (3.26) trivially holds, the lemma has been proven. □

In our case, $0 < \beta < 1$ and $d > 1$, and so the results of Lemma 3.5 and the inequalities in Eq. (3.22) lead to

$$h_l \le \alpha \frac{d^{\beta-1}}{2} (1-\beta)^{\frac{\beta-1}{2\beta}} (1+\beta)^{\frac{\beta+1}{2\beta}} < \alpha\sqrt{e} \tag{3.27}$$

$$h_u \ge \alpha \frac{d^{\beta-1}}{2} (1-\beta)^{\frac{\beta-1}{2\beta}} (1+\beta)^{\frac{\beta+1}{2\beta}} > \frac{\alpha}{2}\frac{1}{d}\sqrt{e}. \tag{3.28}$$

Next, utilizing the condition that $d \le 2n$, from Eq. (3.27) and from Eq. (3.28) we get the following boundaries for α:

$$h_l e^{-\frac{1}{2}} < \alpha < 2h_u d e^{-\frac{1}{2}} \le 4nh_u e^{-\frac{1}{2}}.$$

In summary, the following parameter constraints were set in the minimization problem given by Eq. (3.14):

$$0 < \beta < 1$$
$$\frac{n}{2} \le d \le 2n$$
$$h_l e^{-\frac{1}{2}} < \alpha < 4nh_u e^{-\frac{1}{2}}.$$

In the case of our empirical data, h_l and h_u could be set to $h_l = 0.01$ and $h_u = 0.02$, and since $n = 295$, we had the following parameter constraints:

$$0 < \beta < 1$$
$$\frac{295}{2} \le d \le 590$$
$$0.0061 < \alpha < 14.3141.$$

The optimal values α_{opt}, β_{opt} and d_{opt} of the parameters α, β and d, respectively, identified by using the GLOBAL method are $\alpha_{opt} = 0.069240$, $\beta_{opt} = 0.674587$ and $d_{opt} = 304.12$.

The 3-dimensional surfaces of projections of the Mean Squared Error (MSE) function $MSE(\alpha, \beta, d) = S(\alpha, \beta, d)/n$, in which one of the three parameters is always fixed at its optimal value, and the projections of the global minimum of $MSE(\alpha, \beta, d)$ are shown in Fig. 3.4.

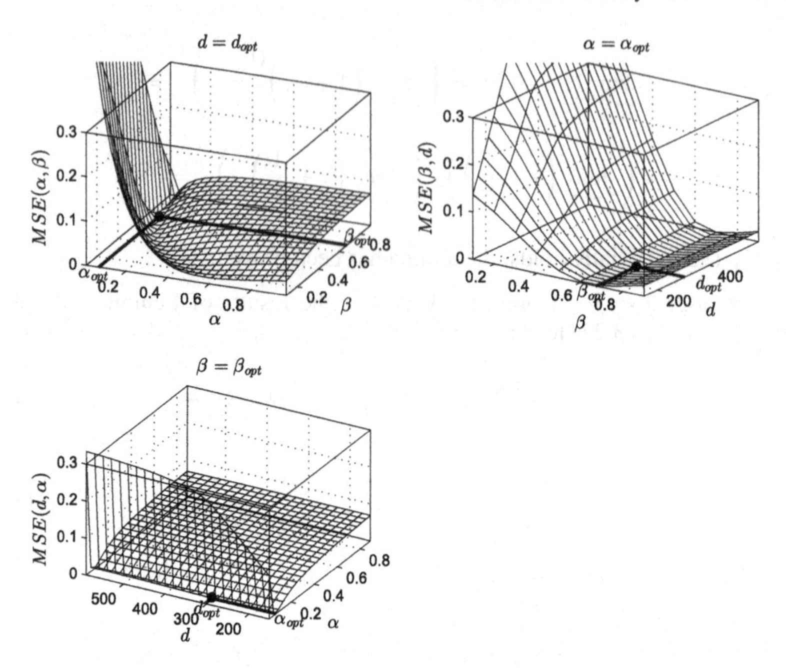

Fig. 3.4 MSE versus parameters of the fitted omega CDF

These 3-dimensional plots can graphically inform us about the sensitivity of the MSE to the values of parameters α, β and d quite well. Utilizing Fig. 3.4, the following general conclusions can be drawn about the sensitivity of function $MSE(\alpha, \beta, d)$. From the right upper and left lower subplots of Fig. 3.4, we can see that the MSE of the fitted omega CDF is much less sensitive to the value of the parameter d than to the values of parameters α and β. At the same time, we can see from the left upper subplot of Fig. 3.4 that at $\alpha = \alpha_{opt}$, $\beta = \beta_{opt}$, $d = d_{opt}$, the function $MSE(\alpha, \beta, d)$ is more sensitive to the value of parameter α than to the value of parameter β.

Our method was compared with the well-known modifications of the Weibull distribution having three parameters. Namely, the cumulative distribution functions of the Modified Weibull Distribution (MW)[19], the Exponentiated Weibull Distribution (EW) [26], the Generalized Weibull Distribution (GWF) [27], the Generalized Power Weibull Distribution (GPW) [28], the Modified Weibull Extension (MWEX) [29], the Odd Weibull Distribution (ODDW) [30], and the Reduced Modified Weibull (RNMW) distribution [31] were also fitted to the same empirical CDF. Tables 3.1 and 3.2 show the cumulative distribution functions and the hazard functions of the examined 3-parameter modifications of the Weibull distribution, respectively. Note that here we apply the same parameter notations as Almalki and Nadarajah did in their review paper [24]. The parameters of each of these distributions were identified like those of the omega distribution; that is, the sum of squared differences between the parametric CDF and the empirical CDF were minimized by using the GLOBAL minimization method. The optimal parameter values and the MSE value for each CDF fitting are summarized in Table 3.3.

Table 3.1 The examined 3-parameter modifications of the Weibul CDF

Model	CDF	Parameter domains
Omega	$F(t)=\begin{cases}1-\left(\frac{d^\beta+t^\beta}{d^\beta-t^\beta}\right)^{\frac{-\alpha d^\beta}{2}}, & \text{if } 0<t<d\\ 1, & \text{if } t\geq d.\end{cases}$	$\alpha,\beta,d>0$
MW	$F(t)=1-\mathrm{e}^{-\beta t^\gamma \mathrm{e}^{\lambda t}}$	$\beta>0,\gamma,\lambda\geq 0$
EW	$F(t)=\left(1-\mathrm{e}^{-\alpha t^\theta}\right)^\lambda$	$\alpha,\theta,\lambda>0$
GWF	$F(t)=1-\left(1-\alpha\lambda t^\theta\right)^{1/\lambda}$	$\alpha,\theta>0,-\infty<\lambda<\infty$
GPW	$F(t)=1-\mathrm{e}^{1-(1+\alpha t^\theta)^{1/\lambda}}$	$\alpha,\theta,\lambda>0$
MWEX	$F(t)=1-\mathrm{e}^{\lambda\alpha^{-1/\theta}\left(1-\mathrm{e}^{\alpha t^\theta}\right)}$	$\alpha,\theta,\lambda>0$
ODDW	$F(t)=1-\left[1+\left(\mathrm{e}^{\alpha t^\theta}-1\right)^\lambda\right]^{-1}$	$\alpha,\theta,\lambda>0$
RNMW	$F(t)=1-\mathrm{e}^{-\alpha\sqrt{t}-\beta\sqrt{t}\mathrm{e}^{\lambda t}}$	$\alpha,\beta,\lambda>0$

$t>0$; for the GWF model if $\lambda>0$, then $t\in(0,(\alpha\lambda)^{-1/\theta})$

Table 3.2 Hazard functions of the examined 3-parameter modifications of the Weibul Distribution

Model	Hazard function	Restriction of parameters if $h(t)$ is bathtub-shaped
Omega	$h(t)=\alpha\beta t^{\beta-1}\frac{d^{2\beta}}{d^{2\beta}-t^{2\beta}}$	$0<\beta<1$
MW	$h(t)=\beta(\gamma+\lambda t)t^{\gamma-1}\mathrm{e}^{\lambda t}$	$0<\theta<1$
EW	$h(t)=\alpha\theta\lambda t^{\theta-1}\mathrm{e}^{-\alpha t^\theta}(1-\mathrm{e}^{-\alpha t^\theta})^{-1}$	$\theta>1;\theta\lambda<1$
GWF	$h(t)=\alpha\theta t^{\theta-1}(1-\alpha\lambda t^\theta)^{-1}$	$\theta<1;0<\lambda$
GPW	$h(t)=\alpha\theta\lambda^{-1}(1+\alpha t^\theta)^{1/\lambda-1}$	$0<\lambda<\theta<1$
MWEX	$h(t)=\lambda\alpha^{((\theta-1)/\theta)}t^{\theta-1}\mathrm{e}^{\alpha t^\theta}$	$\theta<1$
ODDW	$h(t)=\alpha\theta\lambda t^{\theta-1}\mathrm{e}^{\alpha t^\theta}(\mathrm{e}^{\alpha t^\theta}-1)^{\lambda-1}(1+(\mathrm{e}^{\alpha t^\theta}-1)^\lambda)^{-1}$	$\theta>1;\theta\lambda<1$
RNMW	$h(t)=\frac{1}{2\sqrt{t}}(\alpha+\beta(1+2\lambda t)\mathrm{e}^{\lambda t})$	

$t>0$; for the GWF model if $\lambda>0$, then $t\in(0,(\alpha\lambda)^{-1/\theta})$

Table 3.3 CDF fitting results

Function	Parameters			MSE
Omega	$\alpha = 0.069240$	$\beta = 0.674587$	$d = 304.121895$	3.22703e-05
MW	$\beta = 0.078857$	$\gamma = 0.618880$	$\lambda = 2.153\text{e-}03$	3.86417e-05
EW	$\alpha = 2.045\text{e-}03$	$\theta = 1.326312$	$\lambda = 0.396102$	4.78888e-05
GWF	$\alpha = 0.077903$	$\theta = 0.607756$	$\lambda = 0.343967$	3.34385e-05
GPW	$\alpha = 1.33\text{e-}04$	$\theta = 0.534145$	$\lambda = 1.571\text{e-}03$	4.50262e-05
MWEX	$\alpha = 0.072205$	$\theta = 0.550252$	$\lambda = 9.651\text{e-}03$	4.45713e-05
ODDW	$\alpha = 0.020835$	$\theta = 1.027860$	$\lambda = 0.657597$	4.57777e-05
RNMW	$\alpha = 1\text{e-}05$	$\beta = 0.113847$	$\lambda = 3.727\text{e-}03$	1.25777e-04

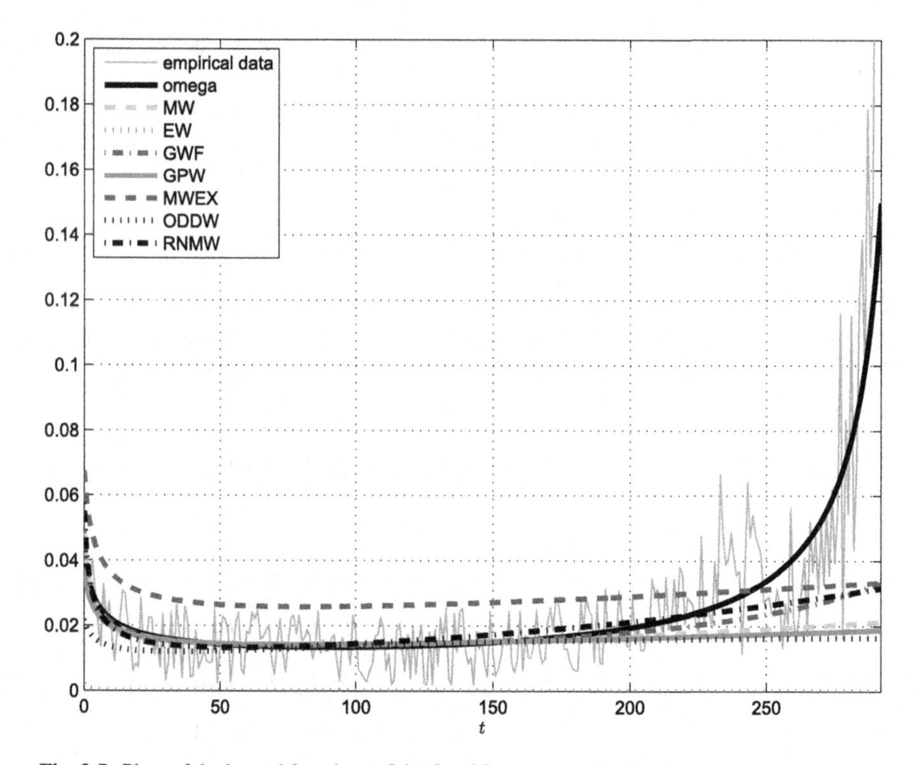

Fig. 3.5 Plots of the hazard functions of the fitted 3-parameter distributions

Figure 3.5 shows the plot of the empirical hazard function $\hat{h}(t)$ and the hazard function plots of the given distributions. Furthermore, for each of the CDFs fitted, the 3-dimensional surfaces of the projections of the MSE of fit as a function of the model parameters, and the optimal model parameter values are shown in Figs. 3.11, 3.12, 3.13, 3.14, 3.15, 3.16 and 3.17 in Appendix 3.5. These 3-dimensional plots express graphically the sensitivity of the MSE to the model parameters quite well.

3.2.2 Approximation to the Exponential Probability Distribution Function

If the random variable η has an exponential probability distribution with the parameter $\alpha > 0$, then the probability distribution function $F_{exp}(x; \alpha)$ of η is given by [6]

$$F_{exp}(x; \alpha) = \begin{cases} 0, & \text{if } x \le 0 \\ 1 - \mathrm{e}^{-\alpha x}, & \text{if } x > 0. \end{cases} \tag{3.29}$$

We will use the notation $\eta \sim \exp(\alpha)$ to indicate that the random variable η has an exponential probability distribution with the parameter $\alpha > 0$. The probability distribution function $F_{exp}(x; \alpha)$ is a special case of the Weibull probability distribution function $F_W(x; \alpha, \beta)$. Namely, if $\beta = 1$, then $F_{exp}(x; \alpha) = F_W(x; \alpha, \beta)$. Based on this, we can state the following theorem [1].

Theorem 3.3 *If $\xi \sim D_p(\alpha, \beta, \gamma, d)$, $\eta \sim \exp(\alpha)$, $\gamma = 1$ and $\beta = 1$, then for any $x \in \mathbb{R}$,*

$$\lim_{d \to \infty} P(\xi < x) = P(\eta < x),$$

where $\alpha, d \in \mathbb{R}$, $\alpha, d > 0$.

Proof The theorem follows from Theorem 3.2. □

This result tells us that if $\gamma = 1, \beta = 1$, then the asymptotic pliant probability distribution function, for the parameter d, is just the exponential probability distribution function.

3.2.2.1 Applications in Reliability Theory

The exponential probability distribution plays a significant role in the theory and practice of reliability management [32, 33]. This distribution also appears frequently in lifetime and reaction time studies. Here, we will discuss how the pliant probability distribution, as an alternative to the exponential distribution, can be applied to model failure rates.

Since the exponential probability distribution may be viewed as a special case of the Weibull probability distribution, namely when $\beta = 1$, utilizing the Weibull hazard function in Eq. (3.16) with $\beta = 1$ gives us the hazard function $h_{exp}(t; \alpha)$ of the exponential distribution with the parameter $\alpha > 0$:

$$h_{exp}(t; \alpha) = h_W(t; \alpha, \beta)_{|\beta=1} = \alpha\beta t^{\beta-1}{}_{|\beta=1} = \alpha.$$

That is, if $\tau \sim \exp(\alpha)$, then the hazard function $h_{exp}(t; \alpha)$ of τ is constant with the value α. Based on Lemma 3.3, the pliant hazard function tends to the Weibull hazard function, if $d \to \infty$. Applying this results for the special case when $\beta = 1$, we get the following reduced pliant hazard function

$$h_P(t;\alpha) = h_P(t;\alpha,\beta)_{|\beta=1} = \alpha\beta t^{\beta-1}\frac{d^{2\beta}}{d^{2\beta}-t^{2\beta}}\Big|_{\beta=1} = \alpha\frac{d^2}{d^2-t^2}, \tag{3.30}$$

where $t \in (0, d)$. There are two key properties of the pliant hazard function $h_P(t;\alpha)$ that should be mentioned here. These are:

1. If $t \in (0, d)$ is fixed and $d \to \infty$, then $h_P(t;\alpha) \to \alpha$.
2. $h_P(t;\alpha)$ is monotonously increasing in the interval $(0, d)$.

The practical implications of the above properties of the hazard function $h_P(t;\alpha)$ are as follows. If d is sufficiently large compared to t, then $h_P(t;\alpha)$ is approximately constant with the value α; that is, the function $h_P(t;\alpha)$ can be utilized to model the quasi constant second phase of a bathtub-shaped failure rate curve. As $h_P(t;\alpha)$ is monotonously increasing in the interval $(0, d)$, it can also be viewed as a model of the increasing third phase of a bathtub-shaped hazard curve [1].

3.2.3 Approximation to the Logistic Probability Distribution Function

The 2-parameter probability distribution function $F_{log}(x;\mu,s)$ of the random variable that has a logistic probability distribution is commonly given by

$$F_L(x;\mu,s) = \frac{1}{1+e^{-\frac{x-\mu}{s}}}, \tag{3.31}$$

where $\mu, s \in \mathbb{R}$ and $s > 0$ are the location and scale parameters of the distribution, respectively [34, 35].

By applying the $\alpha = 1/s$ substitution and setting the location parameter μ to zero, Eq. (3.31) may be written as

$$F_L(x;\alpha) = \frac{1}{1+e^{-\alpha x}}, \tag{3.32}$$

where $\alpha \in \mathbb{R}$, $\alpha > 0$. From now on, we will utilize the logistic probability distribution function in the latter form and use the notation $\eta \sim L(\alpha)$ to indicate that the random variable η has a logistic distribution with the parameter $\alpha > 0$; that is, $P(\eta < x) = F_L(x;\alpha)$. The following proposition will demonstrate that the pliant probability distribution function can be utilized for approximating the logistic distribution function $F_{log}(x;\alpha)$.

Theorem 3.4 *If* $\xi \sim D_p(\alpha, \beta, \gamma, d)$, $\eta \sim L(\alpha)$, $\gamma = -1$, *then for any* $x \in \mathbb{R}$,

$$\lim_{d \to \infty} P(\xi < x) = P(\eta < x),$$

where $\alpha, d \in \mathbb{R}$, $\alpha, d > 0$.

Proof Since $\gamma = -1$ and $\beta = 1$, the pliant probability distribution function $F_P(x; \alpha, \beta, \gamma, d)$ may be written as

$$F_P(x; \alpha, \beta, \gamma, d) = \begin{cases} 0, & \text{if } x \leq 0 \\ \frac{1}{1+\omega_d^{(-\alpha,\beta)}(x)}, & \text{if } x \in (-d, d) \\ 1, & \text{if } x \geq d. \end{cases}$$

Let $x \in \mathbb{R}$ be fixed. Now, utilizing the fact that $\beta = 1$ and exploiting Theorem 3.1 gives

$$F_P(x; \alpha, \beta, \gamma, d) = \frac{1}{1 + \omega_d^{(-\alpha,\beta)}(x)} \xrightarrow{d \to \infty} \frac{1}{1 + \mathrm{e}^{-\alpha x}} = F_L(x; \alpha).$$

This result means that

$$\lim_{d \to \infty} P(\xi < x) = P(\eta < x).$$

□

Based on Theorem 3.4, we can state that the pliant probability distribution function $F_P(x; \alpha, \beta, \gamma, d)$ can be utilized for approximating the logistic distribution function $F_L(x; \alpha)$ [3].

Remark 3.1 Utilizing the following modified omega function, which has an additional parameter μ,

$$\omega_d^{(\alpha,\beta,\mu)}(x) = \left(\frac{d^\beta + (x - \mu)^\beta}{d^\beta - (x - \mu)^\beta} \right)^{\frac{\alpha d^\beta}{2}},$$

where $\alpha, d \in \mathbb{R}, d > 0, \beta \in B_\gamma, x \in \left(\frac{d}{2}(\gamma - 1), d\right), \gamma \in \{-1, 1\}$, would allow us to approximate the 2-parameter logistic probability distribution function given by Eq. (3.31). Namely, it can be shown that if $\alpha = 1/s$, $\gamma = -1$ and $\beta = 1$, then for any $x \in \mathbb{R}$

$$\frac{1}{1 + \omega_d^{(-\alpha,\beta,\mu)}(x)} \xrightarrow{d \to \infty} F_L(x; \mu, s),$$

where $s > 0$. The reason for focusing on the approximation of the 1-parameter logistic distribution function $F_L(x; \alpha)$ is that we will utilize it in an approximation to the standard normal probability distribution function.

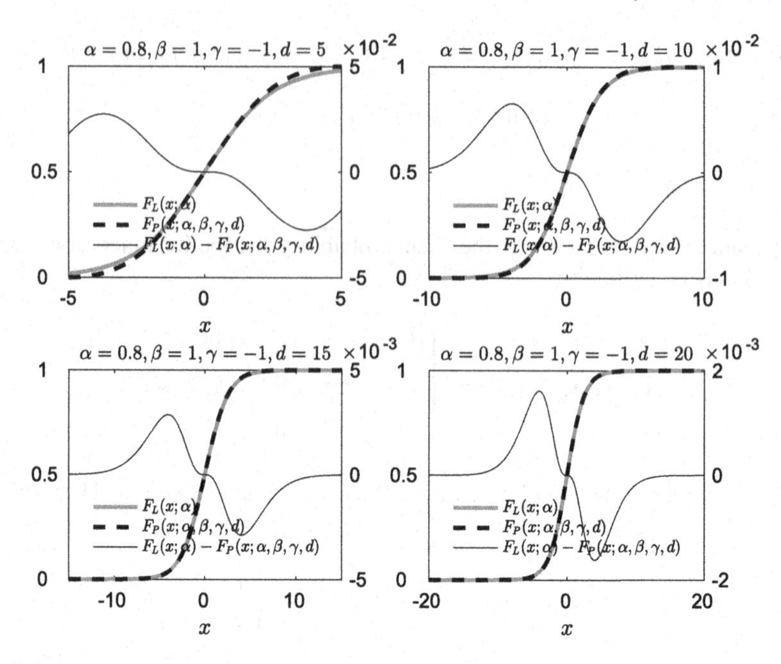

Fig. 3.6 Examples of logistic- and pliant probability distribution function plots

A few sample plots of logistic probability distribution functions and their approximations by pliant probability distribution functions are shown in Fig. 3.6. In each subplot of Fig. 3.6, the left hand side scale is associated with functions $F_L(x; \alpha)$ and $F_P(x; \alpha, \beta, \gamma, d)$, while the right hand side scale is related to the difference function $F_L(x; \alpha) - F_P(x; \alpha, \beta, \gamma, d)$. The plots show how the goodness of approximation improves with increasing values of the parameter d.

3.2.3.1 The Kappa Regression

The logistic regression has a wide range of applications in many fields, including economics, business, biology, the medical sciences and engineering. See some of the recent applications of logistic regression in [36–40]. Articles [41–44] discussed comprehensively the theoretical background of the logistic regression and mentioned a few of its applications. Here, we will demonstrate that a function which can be derived from the pliant probability distribution function may be viewed as an alternative to the logistic regression function [4].

If $\gamma = -1$, then $\beta = 1$ and the formula of the pliant probability distribution function for $x \in (-d, d)$ gives the following $\kappa_d^{(\alpha)}(x)$ function:

$$\kappa_d^{(\alpha)}(x) = \frac{1}{1 + \left(\frac{d+x}{d-x}\right)^{\frac{-\alpha d}{2}}},$$

where $d > 0$ and $\alpha \in \mathbb{R}$. Notice that here we allow α to take any value.

In fuzzy logic, the linguistic modifiers like "very", "more or less", "somewhat", "rather" and "quite" over fuzzy sets that have strictly monotonously increasing or decreasing membership functions, can be modeled by the following unary operator, which is also known as the kappa function [5].

Definition 3.5 The kappa modifier operator (kappa function) is given by

$$\kappa_{\nu,\nu_0}^{(\lambda)}(x) = \frac{1}{1 + \frac{1-\nu_0}{\nu_0}\left(\frac{\nu}{1-\nu}\frac{1-x}{x}\right)^{\lambda}}, \tag{3.33}$$

where $\nu, \nu_0 \in (0, 1)$, $\lambda \in \mathbb{R}$, x is a continuous-valued logic variable.

Now, if we set $\nu = \nu_0 = 0.5$, then the function $\kappa_d^{(\alpha)}(x)$ can be derived from the kappa function $\kappa_{\nu,\nu_0}^{(\lambda)}(x)$ in Eq. (3.33) by setting $\lambda = \alpha d/2$ and applying the $x' = (x + d)/(2d)$ linear transformation ($d > 0$). Next, if $\nu_0 = 0.5$, then the linear transformation of the function $\kappa_{\nu,\nu_0}^{(\lambda)}(x)$ from the domain $(0, 1)$ to the domain (a, b) gives us the following $\kappa(x; \lambda, a, b, x_\nu)$ function:

$$\kappa(x; \lambda, a, b, x_\nu) = \frac{1}{1 + \left(\frac{x_\nu - a}{b - x_\nu}\frac{b-x}{x-a}\right)^{\lambda}}, \tag{3.34}$$

where $\lambda, a, b, x_\nu \in \mathbb{R}$, $x_\nu \in (a, b)$. By using the substitution

$$c = \lambda \ln \frac{x_\nu - a}{b - x_\nu},$$

Equation (3.34) gives the following $\kappa(x; \lambda, a, b, c)$ function:

$$\kappa(x; \lambda, a, b, c) = \frac{1}{1 + e^c\left(\frac{b-x}{x-a}\right)^{\lambda}}, \tag{3.35}$$

where $x \in (a, b)$ and $c, \lambda \in \mathbb{R}$.

Let Y be a dichotomous random variable, and let 0 and 1 code its possible values. Here, we model the conditional probability $P(Y = k|x)$ as a function of the independent variable x; $x \in (a, b)$, $k \in \{0, 1\}$. In our model, which we call the kappa regression model, the odds $O(x)$

$$O(x) = \frac{P(Y = 1|x)}{1 - P(Y = 1|x)} \tag{3.36}$$

are given by

$$\frac{P(Y = 1|x)}{1 - P(Y = 1|x)} = C'\left(\frac{x - a}{b - x}\right)^{\lambda}, \tag{3.37}$$

where $\lambda \in \mathbb{R}$ and $C' > 0$. If C' is written in the form

$$C' = \mathrm{e}^{-c},$$

($c \in \mathbb{R}$) then from Eq. (3.37):

$$P(Y = 1|x) = \frac{1}{1 + \mathrm{e}^{c}\left(\frac{b-x}{x-a}\right)^{\lambda}} = \kappa(x; \lambda, a, b, c). \tag{3.38}$$

In the logistic regression model, the odds in Eq. (3.36) are modeled by

$$O(x) = \frac{P(Y = 1|x)}{1 - P(Y = 1|x)} = \mathrm{e}^{\beta_1 x + \beta_0}, \tag{3.39}$$

and so

$$P(Y = 1|x) = \frac{1}{1 + \mathrm{e}^{-\beta_1 x - \beta_0}}, \tag{3.40}$$

where $\beta_0, \beta_1 \in \mathbb{R}$. The next theorem lays the foundations for using kappa regression as an alternative to logistic regression.

Theorem 3.5 *If $a < b$, $(a + b)/2 = x_0$ is constant,*

$$\lambda = \beta_1 \frac{(x_0 - a)(b - x_0)}{b - a}$$

and $a \to -\infty$, $b \to +\infty$ so that $(a + b)/2 = x_0$, then for any $x \in (a, b)$,

$$C' \left(\frac{x - a}{b - x}\right)^{\lambda} \to \mathrm{e}^{\beta_1 x + \beta_0},$$

where $\beta_0 = \ln C' - \beta_1 x_0$, $(C' > 0)$.

Proof Here, $a \to -\infty$, $b \to +\infty$ so that $(a + b)/2 = x_0$; that is, $a \to -\infty$ and $b \to +\infty$ may be written as $a = x_0 - \Delta$, $b = x_0 + \Delta$, where $\Delta \to \infty$. Then, if the conditions of the lemma are satisfied,

$$\lim_{\substack{a \to -\infty \\ b \to +\infty}} \left(C' \left(\frac{x - a}{b - x}\right)^{\lambda} \right) =$$

$$= \mathrm{e}^{\beta_1 x_0 + \beta_0} \lim_{\substack{a \to -\infty \\ b \to +\infty}} \left(\frac{x - a}{b - x}\right)^{\beta_1 \frac{(x_0 - a)(b - x_0)}{b - a}} = \tag{3.41}$$

$$= \mathrm{e}^{\beta_1 x_0 + \beta_0} \lim_{\Delta \to \infty} \left(\frac{x - x_0 + \Delta}{\Delta - (x - x_0)}\right)^{\beta_1 \frac{\Delta}{2}}.$$

Table 3.4 Data from the heat endurance experiment

r	1	2	3	4	5	6	7	8	9	10	11
t_r [C°]	250	252	254	256	258	260	262	264	266	268	270
k_r	0	5	3	13	12	29	33	55	78	97	100
$n_r - k_r$	100	95	97	87	88	71	67	45	22	3	0

Similar to the proof of Theorem 3.1, it can be shown that

$$\lim_{\Delta\to\infty}\left(\frac{x-x_0+\Delta}{\Delta-(x-x_0)}\right)^{\beta_1\frac{\Delta}{2}} = e^{\beta_1(x-x_0)},$$

and so the chain in Eq. (3.41) can be continued as

$$e^{\beta_1 x_0+\beta_0}\lim_{\Delta\to\infty}\left(\frac{x-x_0+\Delta}{\Delta-(x-x_0)}\right)^{\beta_1\frac{\Delta}{2}} = \\ = e^{\beta_1 x_0+\beta_0}e^{\beta_1(x-x_0)} = e^{\beta_1 x+\beta_0}.$$

□

Based on Theorem 3.5, logistic regression may be viewed as the asymptotic kappa regression. It is worth adding that in the case of logistic regression, the odds $O(x)$ take all the positive real numbers in the unbounded domain $(-\infty, +\infty)$, while in the kappa regression model, the odds $O(x)$ take all the positive real values in the bounded domain (a, b). That is, the kappa regression may be treated as a quasi logistic regression, where the explanatory variable is defined over a bounded subset of the real numbers. This property of the kappa regression is advantageous in situations where conditional likelihoods close to zero or close to one need to be modeled in a bounded interval.

A Demonstrative Example

The reliability of an electronic component was investigated in a heat stress experiment. Altogether 1100 components were individually heat stressed so that at each designated temperature between 250 and 270 °C, 100 components were individually heated for 20 s. After the heat stress, each component was functionally tested and the number of components that failed the test (k_r) as well as the number of those that passed it ($n_r - k_r$) out of the $n_r = 100$ components were counted. The results are summarized in Table 3.4. The purpose of the experiment was to model the probability that a component of the examined type fails at a given value of the heat stress. Let the dichotomous random variable Y be 1, if a component fails, and let it be 0, if it passes the heat stress test. With this notation, our goal was to model the conditional probability $P(Y = 1|t)$, where t denotes the temperature in degrees Celsius, $t \in [250, 270]$.

Table 3.5 Regression results

Model	β_0	β_1	c	λ	$-l(\Theta)$	R
Logistic	−104.3770	0.3975	–	–	386.7473	0.9878
Kappa	–	–	0.9909	1.5510	376.5254	0.9945

The experimental data in Table 3.4 were used to build a logistic regression and a kappa regression model. Using the experimental data, the log-likelihood function may be written as

$$l(\Theta) = \sum_{r=1}^{11} k_r \ln p(t_r; \Theta) + \sum_{r=1}^{11} (n_r - k_r) \ln(1 - p(t_r; \Theta)). \tag{3.42}$$

For logistic regression $\Theta = (\beta_0, \beta_1)$ and

$$p(t_r; \Theta) = \sigma(t_r; \beta_0, \beta_1) = \frac{1}{1 + e^{-\beta_1 t_r - \beta_0}}. \tag{3.43}$$

For kappa regression $\Theta = (c, \lambda)$ and

$$p(t_r; \Theta) = \kappa(t_r; c, \lambda) = \frac{1}{1 + e^c \left(\frac{b - t_r}{t_r - a}\right)^\lambda}. \tag{3.44}$$

As the kappa function in Eq. (3.44) is defined in the open interval (a, b) and the temperature in the experiment was in the interval [250, 270], parameters a and b were set to $a = 250 - \delta$, $b = 270 + \delta$, where δ is a small positive number. In our implementation $\delta = 10^{-3}$. In both regression approaches, the maximum of the log-likelihood function was numerically computed by applying the gradient descent method to the negative log-likelihood function. The results of the two regression models, including the correlation coefficient R as well, are summarized in Table 3.5. Figure 3.7 shows the convex surfaces of the negative log-likelihood functions $-l(\beta_0, \beta_1)$ and $-l(c, \lambda)$ for logistic regression and kappa regression, respectively. Here, $\beta_{0,opt}$ and $\beta_{1,opt}$ denote, respectively, the values of β_0 and β_1 that minimize $-l(\beta_0, \beta_1)$, while c_{opt} and λ_{opt} denote, respectively, the values of c and λ at which $-l(c, \lambda)$ is minimal. Figure 3.8 shows how the logistic regression model and kappa regression model match the empirical conditional likelihoods

$$P(Y = 1|t_r) = \frac{k_r}{n_r} \tag{3.45}$$

$(r = 1, 2, \ldots, 11)$, which are computed from the sample. The statistical significance of parameter λ of the kappa regression function was tested using a likelihood ratio test method (see [4]), which resulted in a p-value of 0. Hence, the parameter λ is statistically significant.

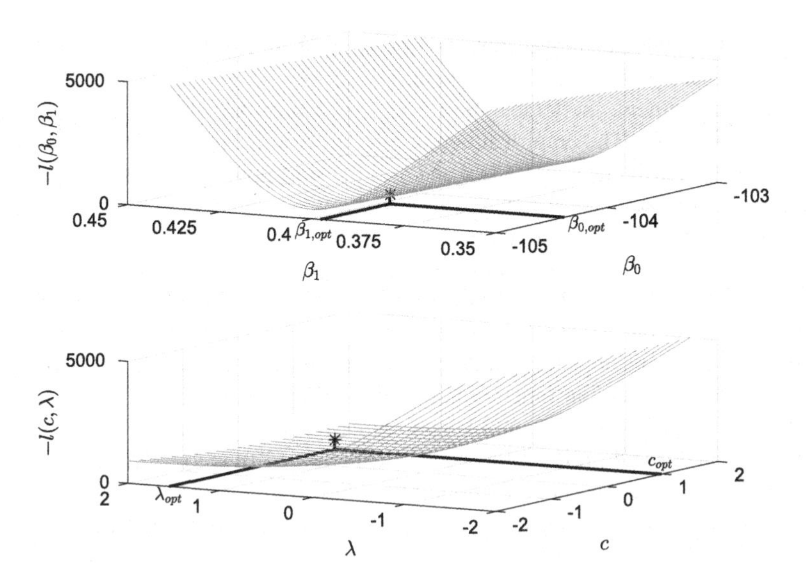

Fig. 3.7 The negative log-likelihood functions

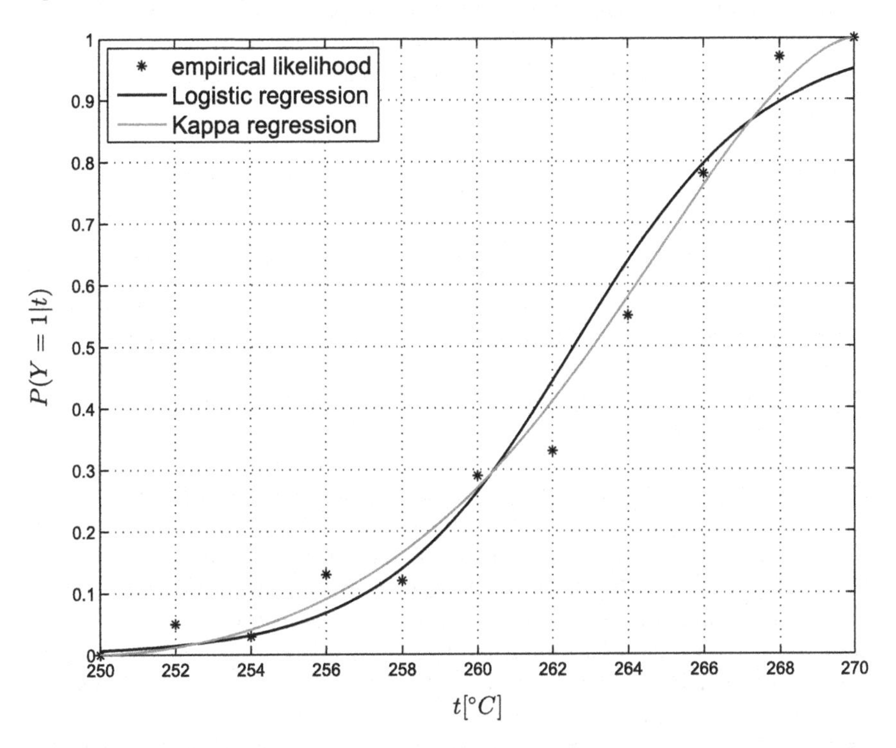

Fig. 3.8 Curves of the logistic and kappa regression functions

Noting the above example, there are a couple of important properties of kappa regression that should be mentioned here.

In the experiment, the temperature t, as explanatory variable, is defined in the interval $[250, 270]$. The empirical conditional probability of the event that a component fails after the heat stress is 0 and 1 at the start point and at the end point of domain of t, respectively. An increasing logistic function neither takes the value of 0, nor the value of 1; these are its limits as the independent variable goes to $-\infty$ or $+\infty$, respectively. Hence, there might be cases where in a given interval the logistic function cannot properly approximate empirical conditional likelihoods that have the values of zero or one, or are very close to these values. In our experiment, based on the sample data, the empirical conditional likelihood $P(Y = 1|t = 270)$ is 1, while the the logistic regression function has a value of approximately 0.95 at $t = 270$. That is, although the logistic and kappa regression functions fit quite well to the empirical conditional probabilities and their likelihood function values are similar as well, in this case the kappa function fits to the empirical data better at the right end of the domain of t. This means that kappa regression has a slightly greater log-likelihood value than logistic regression does. Based on this, we may conclude that in cases where the domain of the explanatory variable is an interval and the empirical conditional probabilities are very close to zero or one at the end points of the domain, kappa regression may give better results than logistic regression.

The endpoints of interval (a, b) for the kappa regression were set to

$$\begin{aligned} a &= \min_{r=1,\ldots,11} (t_r) - \delta \\ b &= \max_{r=1,\ldots,11} (t_r) + \delta. \end{aligned} \tag{3.46}$$

In our study, δ was set to $\delta = 10^{-3}$. Here, Table 3.6 shows how the parameter δ affects the regression parameters c and λ, and the minimum value of the negative log-likelihood function. We can see from Table 3.6 that although the minimum of the negative log-likelihood function slightly decreases with δ, and δ has a small effect on c and λ, if $\delta \leq 10^{-3}$, decreasing δ further has a negligible effect on the regression results.

Noting Theorem 3.5, if $a \to -\infty$ and $b \to +\infty$, then the kappa function asymptotically tends to the logistic function. Kappa regression on the empirical data of the above example was computed with various parameter pairs (a, b), where $(a + b)/2 = 260$. The minimum value of the negative log-likelihood function of kappa regression was computed for each parameter pair (a, b), the results of which are shown in Table 3.7. We observe from these results that as the width of interval (a, b) increases, the value of the negative log-likelihood function increases and gets closer to that of logistic regression. From Fig. 3.9, we can see how the kappa regression curve is getting closer to the logistic regression curve as a decreases and b increases. This behavior of the kappa regression function is in line with the results of Theorem 3.5.

Table 3.6 Impact of δ

δ	c	λ	$-l(c, \lambda)$
10^{-1}	0.9926	1.5782	376.6569
10^{-2}	0.9910	1.5532	376.5330
10^{-3}	0.9909	1.5510	376.5254
10^{-4}	0.9909	1.5507	376.5248
10^{-5}	0.9909	1.5507	376.5247
10^{-6}	0.9909	1.5507	376.5247
10^{-7}	0.9909	1.5507	376.5247
10^{-8}	0.9909	1.5507	376.5247

Table 3.7 Impact of a and b ($\delta = 10^{-3}$)

a	b	$-l(\beta_0, \beta_1)$	$-l(c, \lambda)$
$250 - \delta$	$270 + \delta$	386.7473	376.5254
$245 - \delta$	$275 + \delta$	386.7473	382.6631
$240 - \delta$	$280 + \delta$	386.7473	384.5545
$235 - \delta$	$285 + \delta$	386.7473	385.3724

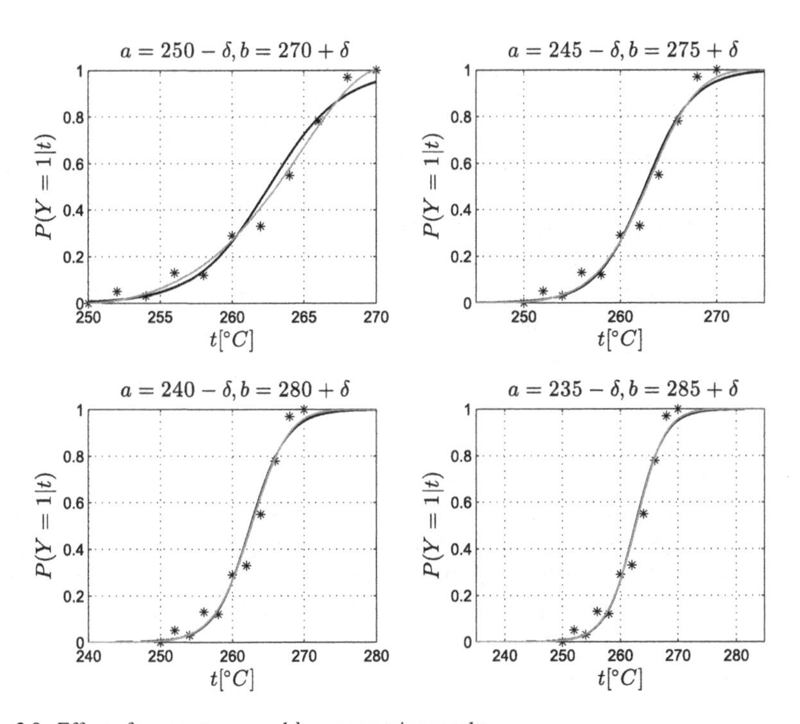

Fig. 3.9 Effect of parameters a and b on regression results

The findings from the above example can be summarized as follows. Kappa regression is applicable when binary logistic regression is applicable, and these two methods give similar results. Logistic regression does not provide the most appropriate mathematical model if point symmetry is lacking or if the predictor variables are confined to finite intervals (a, b) of real numbers. That is, if the explanatory variable is defined over a bounded subset of real numbers, the empirical conditional probabilities exhibit an asymmetric trend and there are empirical conditional probabilities very close to zero or one at the terminal locations of the domain of the explanatory variable, then kappa regression provides the proper mathematical model.

3.2.4 Approximation to the Standard Normal Probability Distribution Function

The probability distribution function $\Phi(x)$ of the standard normal random variable is given by

$$\Phi(x) = \frac{1}{\sqrt{2\pi}} \int_{-\infty}^{x} \mathrm{e}^{-\frac{t^2}{2}} \mathrm{d}t.$$

We will use the common notation $\eta \sim N(0, 1)$ to indicate that the random variable η has the standard normal probability distribution. The fact that the probability distribution function $\Phi(x)$ cannot be expressed in a closed form and the practical needs for computing its values provided the motivations for researchers and practitioners to approximate the standard normal probability distribution function. These research efforts resulted in an extremely wide range of approximations with various applications. A considerable portion of the approximations is the group of ad-hoc methods which typically utilize an a priori selected parametric function and apply various mathematical techniques to estimate the parameters in order minimize the approximation error. The authors of [45–47] gave comprehensive overviews of the approximation formulas in their reviews. In general, we may state that the accuracy of approximations increases with the complexity of formulas and with the number of parameters they have.

Among the many ad-hoc approximations, the 1-parameter logistic function, which has the same form as the logistic probability distribution function $F_L(x; \alpha)$, can also be applied to approximate the standard normal probability distribution function (see, for example, the books of [35, 48, 49], and the paper of [50]). Here, we will utilize Tocher's approximation and the pliant probability distribution function to obtain a novel approximation (which has only one parameter and a very simple formula) to the standard normal probability distribution function.

Tocher's approximation, which we denote by $\Phi_T(x)$, utilizes the logistic probability distribution function $F_L(x;\alpha)$ in Eq. (3.32) with $\alpha = 2\sqrt{2/\pi}$. That is,

$$\Phi_T(x) = \frac{1}{1+\mathrm{e}^{-2\sqrt{2/\pi}x}}. \tag{3.47}$$

It is worth mentioning here that the probability density function of Tocher's approximation can be derived from sigmoid fuzzy membership functions by using operators of continuous-valued logic [3]. Note that setting α to $2\sqrt{2/\pi}$ ensures that function $\Phi_T(x)$ is identical with function $\Phi(x)$ to first order at $x = 0$. Furthermore, based on Theorem 3.4, if $\alpha = 2\sqrt{2/\pi}$, $\gamma = -1$ and $\beta = 1$, then for any $x \in \mathbb{R}$,

$$\lim_{d\to\infty} F_P(x;\alpha,\beta,\gamma,d) = \Phi_T(x).$$

These results tell us that the following function, which we call the quasi logistic probability distribution function

$$\Phi_{\kappa,d}(x) = \begin{cases} 0, & \text{if } x \le 0 \\ \frac{1}{1+\left(\frac{d+x}{d-x}\right)^{-\sqrt{2/\pi}d}}, & \text{if } x \in (-d,d) \\ 1, & \text{if } x \ge d \end{cases}$$

can be applied to approximate the standard normal probability distribution function $\Phi(x)$, where $d > 0$. Notice that $\Phi_{\kappa,d}(x) = F_P(x;\alpha,\beta,\gamma,d)$ with the parameter values $\alpha = 2\sqrt{2/\pi}, \gamma = -1$ and $\beta = 1$, so the function $\Phi_{\kappa,d}(x)$ is in fact a probability distribution function. Furthermore, if $\nu = \nu_0 = 0.5$, then the function $\Phi_{\kappa,d}(x)$ for $x \in (-d,d)$ can be derived from the kappa function $\kappa_{\nu,\nu_0}^{(\lambda)}(x)$ in Eq. (3.33) by setting $\lambda = \sqrt{2/\pi}d$ and applying the $x' = (x+d)/(2d)$ linear transformation ($d > 0$). The κ index in the notation $\Phi_{\kappa,d}(x)$ indicates that this function is connected with the kappa modifier operator.

It can be shown numerically that $|\Phi(x) - \Phi_{\kappa,d}(x)|$ is approximately minimal, if $d = 3.1152$. In this case, the maximum absolute approximation error is $2.15 \cdot 10^{-3}$. Considering that 3.1152 is close to π, using $d = \pi$ instead of $d = 3.1152$ does not worsen significantly the approximation accuracy. If $d = \pi$, then the maximum absolute approximation error is $2.3570 \cdot 10^{-3}$. Thus, we propose the use of function $\Phi_{\kappa,\pi}(x)$ for approximation as this function has a very simple formula and its maximum absolute approximation error is just slightly greater than that of function $\Phi_{\kappa,d}(x)$ with $d = 3.1152$ [3]:

$$\Phi_{\kappa,\pi}(x) = \Phi_{\kappa,d}(x)\Big|_{d=\pi} = \begin{cases} 0, & \text{if } x \le -\pi \\ \frac{1}{1+\left(\frac{\pi-x}{\pi+x}\right)^{\sqrt{2\pi}}}, & \text{if } x \in (-\pi,+\pi) \\ 1, & \text{if } x \ge +\pi. \end{cases}$$

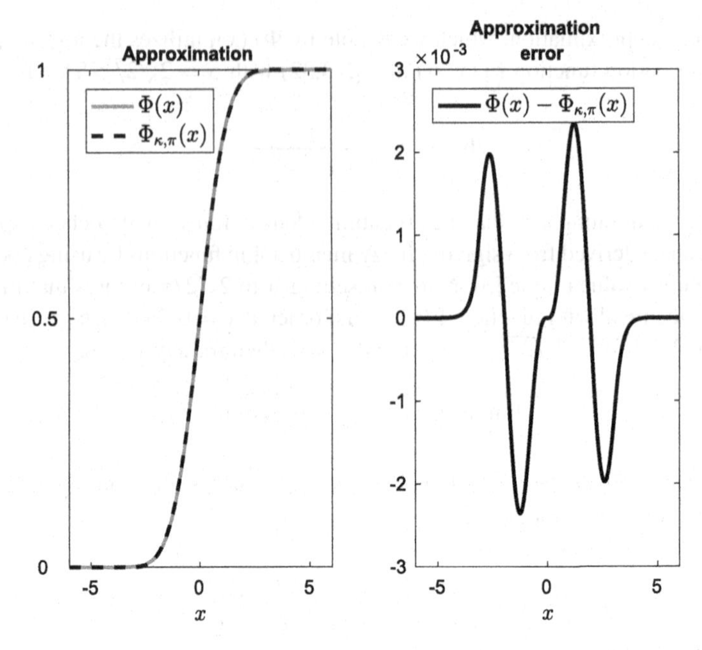

Fig. 3.10 Approximation to the standard normal distribution function by a pliant (quasi logistic) distribution function

Recall that $\Phi_{\kappa,\pi}(x) = F_P(x; \alpha, \beta, \gamma, d)$ with the parameter values $\alpha = 2\sqrt{2/\pi}$, $\beta = 1, \gamma = -1$ and $d = \pi$; that is, $\Phi_{\kappa,\pi}(x)$ is a special case of the pliant probability distribution function. It should be added that there are just a few known approximations with a single constant parameter in this accuracy range (e.g. [51–54]), and all these approximations include exponential terms, while $\Phi_{\kappa,\pi}(x)$ does not contain any. That is, based on our current knowledge, in this accuracy range, there is no other known approximation that has such a simple formula as $\Phi_{\kappa,\pi}(x)$. The known approximations that have a better accuracy have more complex formulas, while the ones with similar complex formulas do not have a higher accuracy.

Figure 3.10 shows that the pliant probability distribution function fits quite well to the standard normal probability distribution function. Based on the above results, we can state the following proposition.

Proposition 3.1 *If* $\xi \sim D_p(\alpha, \beta, \gamma, d), \eta \sim N(0, 1), \alpha = 2\sqrt{2/\pi}, \gamma = -1$ *and* $d = \pi$ *then for any* $x \in \mathbb{R}$,

$$\max_{x \in \mathbb{R}} |P(\eta < x) - P(\xi < x)| < 2.36 \cdot 10^{-3}.$$

Proof See the previous results of this section. □

Table 3.8 Summary of approximations by the pliant CDF

Parameters and domain of $F_P(x) = F_P(x; \alpha, \beta, \gamma, d)$	Approximated CDF	Approximation error
$\alpha > 0, \beta > 0$; $\gamma = 1, d > 0$; $x \in (0, d)$	$F_W(x; \alpha, \beta) = \begin{cases} 0, & x \leq 0 \\ 1 - \exp(-\alpha x^{\beta}), & x > 0 \end{cases}$	$\lim_{d \to \infty} \lvert F_W(x; \alpha, \beta) - F_P(x) \rvert = 0$
$\alpha > 0, \beta = 1$; $\gamma = 1, d > 0$; $x \in (0, d)$	$F_{exp}(x; \alpha) = \begin{cases} 0, & x \leq 0 \\ 1 - \exp(-\alpha x), & x > 0 \end{cases}$	$\lim_{d \to \infty} \lvert F_{exp}(x; \alpha) - F_P(x) \rvert = 0$
$\alpha > 0, \beta = 1$; $\gamma = -1, d > 0$; $x \in (-d, d)$	$F_L(x; \alpha) = \dfrac{1}{1 + \exp(-\alpha x)}$	$\lim_{d \to \infty} \lvert F_L(x; \alpha) - F_P(x) \rvert = 0$
$\alpha = 2\sqrt{2/\pi}, \beta = 1$; $\gamma = -1, d > 0$; $x \in (-d, d)$	$\Phi(x) = \dfrac{1}{\sqrt{2\pi}} \int_{-\infty}^{x} \exp\left(-\dfrac{t^2}{2}\right) \mathrm{d}t$	$\max_{x \in \mathbb{R}} \lvert \Phi(x) - F_P(x) \rvert < 2.36 \cdot 10^{-3}$

3.3 Summary

Table 3.8 summarizes how the pliant probability distribution function (pliant CDF)

$$F_P(x; \alpha, \beta, \gamma, d) = \begin{cases} 0, & \text{if } x \leq \frac{d}{2}(\gamma - 1) \\ \left(1 - \gamma \omega_d^{(-\alpha, \beta)}(x)\right)^{\gamma}, & \text{if } x \in \left(\frac{d}{2}(\gamma - 1), d\right) \\ 1, & \text{if } x \geq d, \end{cases}$$

where $\alpha, d \in \mathbb{R}$, $\alpha > 0$, $d > 0$, $\beta \in B_{\gamma}$, $\gamma \in \{-1, 1\}$, can be applied to approximate some well-known probability distribution functions.

Based on the theoretical results and findings in this chapter, we may conclude that the pliant probability distribution function may be viewed as an alternative to some key probability distribution functions including the Weibull-, exponential, logistic and standard normal probability distribution functions. We showed that our results can be utilized in a wide range of fields that include engineering, economics and the social sciences.

3.4 Appendix: Empirical Data and the Values of Empirical CDF and HF

See Tables 3.9, 3.10 and 3.11.

Table 3.9 Empirical data and the values of empirical CDF and empirical HF for weeks 0–99

i	$N(i)$	$\hat{F}(i)$	$\hat{h}(i)$	i	$N(i)$	$\hat{F}(i)$	$\hat{h}(i)$
0	21000	0.0000	0.0686	50	8018	0.6182	0.0025
1	19559	0.0686	0.0381	51	7998	0.6191	0.0139
2	18814	0.1041	0.0315	52	7887	0.6244	0.0229
3	18221	0.1323	0.0327	53	7706	0.6330	0.0218
4	17626	0.1607	0.0404	54	7538	0.6410	0.0188
5	16914	0.1946	0.0305	55	7396	0.6478	0.0027
6	16398	0.2191	0.0208	56	7376	0.6488	0.0115
7	16057	0.2354	0.0135	57	7291	0.6528	0.0214
8	15841	0.2457	0.0327	58	7135	0.6602	0.0241
9	15323	0.2703	0.0214	59	6963	0.6684	0.0069
10	14995	0.2860	0.0273	60	6915	0.6707	0.0051
11	14585	0.3055	0.0167	61	6880	0.6724	0.0182
12	14341	0.3171	0.0091	62	6755	0.6783	0.0176
13	14210	0.3233	0.0183	63	6636	0.6840	0.0243
14	13950	0.3357	0.0216	64	6475	0.6917	0.0221
15	13649	0.3500	0.0308	65	6332	0.6985	0.0246
16	13229	0.3700	0.0297	66	6176	0.7059	0.0159
17	12836	0.3888	0.0213	67	6078	0.7106	0.0165
18	12563	0.4018	0.0172	68	5978	0.7153	0.0187
19	12347	0.4120	0.0061	69	5866	0.7207	0.0201
20	12272	0.4156	0.0084	70	5748	0.7263	0.0090
21	12169	0.4205	0.0071	71	5696	0.7288	0.0051
22	12082	0.4247	0.0284	72	5667	0.7301	0.0148
23	11739	0.4410	0.0144	73	5583	0.7341	0.0220
24	11570	0.4490	0.0148	74	5460	0.7400	0.0123
25	11399	0.4572	0.0226	75	5393	0.7432	0.0057
26	11141	0.4695	0.0223	76	5362	0.7447	0.0145
27	10893	0.4813	0.0111	77	5284	0.7484	0.0223
28	10772	0.4870	0.0154	78	5166	0.7540	0.0232
29	10606	0.4950	0.0072	79	5046	0.7597	0.0174
30	10530	0.4986	0.0237	80	4958	0.7639	0.0232
31	10280	0.5105	0.0032	81	4843	0.7694	0.0155
32	10247	0.5120	0.0087	82	4768	0.7730	0.0161
33	10158	0.5163	0.0244	83	4691	0.7766	0.0188
34	9910	0.5281	0.0069	84	4603	0.7808	0.0165
35	9842	0.5313	0.0264	85	4527	0.7844	0.0097
36	9582	0.5437	0.0177	86	4483	0.7865	0.0129
37	9412	0.5518	0.0040	87	4425	0.7893	0.0244
38	9374	0.5536	0.0258	88	4317	0.7944	0.0116

(continued)

Table 3.9 (continued)

i	$N(i)$	$\hat{F}(i)$	$\hat{h}(i)$	i	$N(i)$	$\hat{F}(i)$	$\hat{h}(i)$
39	9132	0.5651	0.0219	89	4267	0.7968	0.0155
40	8932	0.5747	0.0041	90	4201	0.8000	0.0121
41	8895	0.5764	0.0046	91	4150	0.8024	0.0123
42	8854	0.5784	0.0185	92	4099	0.8048	0.0161
43	8690	0.5862	0.0067	93	4033	0.8080	0.0174
44	8632	0.5890	0.0093	94	3963	0.8113	0.0093
45	8552	0.5928	0.0124	95	3926	0.8130	0.0031
46	8446	0.5978	0.0091	96	3914	0.8136	0.0164
47	8369	0.6015	0.0086	97	3850	0.8167	0.0135
48	8297	0.6049	0.0088	98	3798	0.8191	0.0140
49	8224	0.6084	0.0250	99	3745	0.8217	0.0080

Table 3.10 Empirical data and the values of empirical CDF and empirical HF for weeks 100–199

i	$N(i)$	$\hat{F}(i)$	$\hat{h}(i)$	i	$N(i)$	$\hat{F}(i)$	$\hat{h}(i)$
100	3715	0.8231	0.0234	150	2007	0.9044	0.0244
101	3628	0.8272	0.0182	151	1958	0.9068	0.0123
102	3562	0.8304	0.0065	152	1934	0.9079	0.0047
103	3539	0.8315	0.0054	153	1925	0.9083	0.0068
104	3520	0.8324	0.0037	154	1912	0.9090	0.0105
105	3507	0.8330	0.0029	155	1892	0.9099	0.0164
106	3497	0.8335	0.0063	156	1861	0.9114	0.0140
107	3475	0.8345	0.0081	157	1835	0.9126	0.0245
108	3447	0.8359	0.0168	158	1790	0.9148	0.0084
109	3389	0.8386	0.0174	159	1775	0.9155	0.0056
110	3330	0.8414	0.0225	160	1765	0.9160	0.0096
111	3255	0.8450	0.0197	161	1748	0.9168	0.0246
112	3191	0.8480	0.0063	162	1705	0.9188	0.0065
113	3171	0.8490	0.0114	163	1694	0.9193	0.0171
114	3135	0.8507	0.0019	164	1665	0.9207	0.0102
115	3129	0.8510	0.0035	165	1648	0.9215	0.0103
116	3118	0.8515	0.0186	166	1631	0.9223	0.0141
117	3060	0.8543	0.0206	167	1608	0.9234	0.0249
118	2997	0.8573	0.0127	168	1568	0.9253	0.0134
119	2959	0.8591	0.0061	169	1547	0.9263	0.0084
120	2941	0.8600	0.0173	170	1534	0.9270	0.0098
121	2890	0.8624	0.0038	171	1519	0.9277	0.0211
122	2879	0.8629	0.0195	172	1487	0.9292	0.0256
123	2823	0.8656	0.0053	173	1449	0.9310	0.0186

(continued)

Table 3.10 (continued)

i	$N(i)$	$\hat{F}(i)$	$\hat{h}(i)$	i	$N(i)$	$\hat{F}(i)$	$\hat{h}(i)$
124	2808	0.8663	0.0153	174	1422	0.9323	0.0288
125	2765	0.8683	0.0203	175	1381	0.9342	0.0261
126	2709	0.8710	0.0022	176	1345	0.9360	0.0201
127	2703	0.8713	0.0018	177	1318	0.9372	0.0129
128	2698	0.8715	0.0182	178	1301	0.9380	0.0131
129	2649	0.8739	0.0019	179	1284	0.9389	0.0093
130	2644	0.8741	0.0189	180	1272	0.9394	0.0071
131	2594	0.8765	0.0077	181	1263	0.9399	0.0063
132	2574	0.8774	0.0225	182	1255	0.9402	0.0064
133	2516	0.8802	0.0151	183	1247	0.9406	0.0273
134	2478	0.8820	0.0254	184	1213	0.9422	0.0190
135	2415	0.8850	0.0037	185	1190	0.9433	0.0286
136	2406	0.8854	0.0021	186	1156	0.9450	0.0251
137	2401	0.8857	0.0237	187	1127	0.9463	0.0231
138	2344	0.8884	0.0119	188	1101	0.9476	0.0091
139	2316	0.8897	0.0246	189	1091	0.9480	0.0275
140	2259	0.8924	0.0252	190	1061	0.9495	0.0085
141	2202	0.8951	0.0127	191	1052	0.9499	0.0086
142	2174	0.8965	0.0083	192	1043	0.9503	0.0316
143	2156	0.8973	0.0102	193	1010	0.9519	0.0158
144	2134	0.8984	0.0019	194	994	0.9527	0.0191
145	2130	0.8986	0.0183	195	975	0.9536	0.0195
146	2091	0.9004	0.0129	196	956	0.9545	0.0146
147	2064	0.9017	0.0155	197	942	0.9551	0.0308
148	2032	0.9032	0.0079	198	913	0.9565	0.0230
149	2016	0.9040	0.0045	199	892	0.9575	0.0179

Table 3.11 Empirical data and the values of empirical CDF and empirical HF for weeks 200–295

i	$N(i)$	$\hat{F}(i)$	$\hat{h}(i)$	i	$N(i)$	$\hat{F}(i)$	$\hat{h}(i)$
200	876	0.9583	0.0137	250	183	0.9913	0.0273
201	864	0.9589	0.0139	251	178	0.9915	0.0281
202	852	0.9594	0.0282	252	173	0.9918	0.0289
203	828	0.9606	0.0145	253	168	0.9920	0.0238
204	816	0.9611	0.0196	254	164	0.9922	0.0244
205	800	0.9619	0.0125	255	160	0.9924	0.0313
206	790	0.9624	0.0127	256	155	0.9926	0.0258
207	780	0.9629	0.0167	257	151	0.9928	0.0132
208	767	0.9635	0.0196	258	149	0.9929	0.0268

(continued)

Table 3.11 (continued)

i	$N(i)$	$\hat{F}(i)$	$\hat{h}(i)$	i	$N(i)$	$\hat{F}(i)$	$\hat{h}(i)$
209	752	0.9642	0.0346	259	145	0.9931	0.0207
210	726	0.9654	0.0303	260	142	0.9932	0.0563
211	704	0.9665	0.0313	261	134	0.9936	0.0224
212	682	0.9675	0.0205	262	131	0.9938	0.0153
213	668	0.9682	0.0180	263	129	0.9939	0.0233
214	656	0.9688	0.0351	264	126	0.9940	0.0317
215	633	0.9699	0.0190	265	122	0.9942	0.0410
216	621	0.9704	0.0209	266	117	0.9944	0.0171
217	608	0.9710	0.0345	267	115	0.9945	0.0522
218	587	0.9720	0.0204	268	109	0.9948	0.0367
219	575	0.9726	0.0383	269	105	0.9950	0.0476
220	553	0.9737	0.0362	270	100	0.9952	0.0300
221	533	0.9746	0.0206	271	97	0.9954	0.0515
222	522	0.9751	0.0287	272	92	0.9956	0.0326
223	507	0.9759	0.0276	273	89	0.9958	0.0562
224	493	0.9765	0.0264	274	84	0.9960	0.0357
225	480	0.9771	0.0208	275	81	0.9961	0.0617
226	470	0.9776	0.0170	276	76	0.9964	0.0395
227	462	0.9780	0.0476	277	73	0.9965	0.0548
228	440	0.9790	0.0205	278	69	0.9967	0.1159
229	431	0.9795	0.0255	279	61	0.9971	0.0164
230	420	0.9800	0.0238	280	60	0.9971	0.0833
231	410	0.9805	0.0195	281	55	0.9974	0.0545
232	402	0.9809	0.0398	282	52	0.9975	0.1154
233	386	0.9816	0.0259	283	46	0.9978	0.0435
234	376	0.9821	0.0665	284	44	0.9979	0.0682
235	351	0.9833	0.0484	285	41	0.9980	0.1220
236	334	0.9841	0.0389	286	36	0.9983	0.1389
237	321	0.9847	0.0405	287	31	0.9985	0.0968
238	308	0.9853	0.0487	288	28	0.9987	0.1786
239	293	0.9860	0.0410	289	23	0.9989	0.1304
240	281	0.9866	0.0356	290	20	0.9990	0.1500
241	271	0.9871	0.0369	291	17	0.9992	0.2941
242	261	0.9876	0.0115	292	12	0.9994	0.4167
243	258	0.9877	0.0349	293	7	0.9997	0.4286
244	249	0.9881	0.0643	294	4	0.9998	1.0000
245	233	0.9889	0.0472	295	0	1.0000	
246	222	0.9894	0.0541				
247	210	0.9900	0.0476				
248	200	0.9905	0.0450				
249	191	0.9909	0.0419				

3.5 Appendix: MSE of Fit Versus Parameter Values of the Fitted CDF

See Figs. 3.11, 3.12, 3.13, 3.14, 3.15, 3.16 and 3.17.

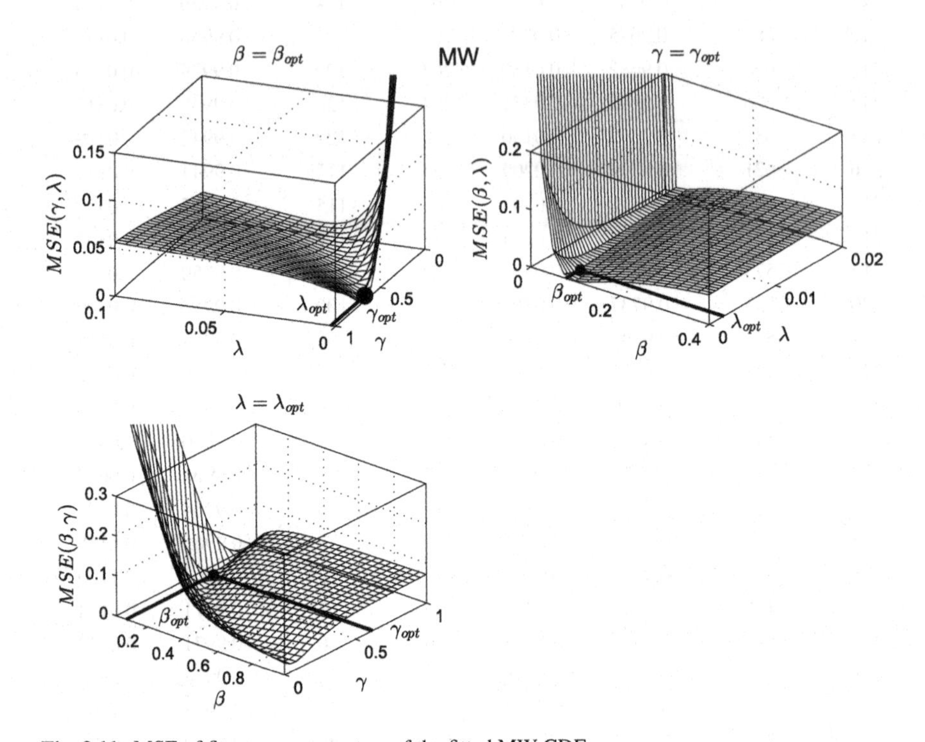

Fig. 3.11 MSE of fit versus parameters of the fitted MW CDF

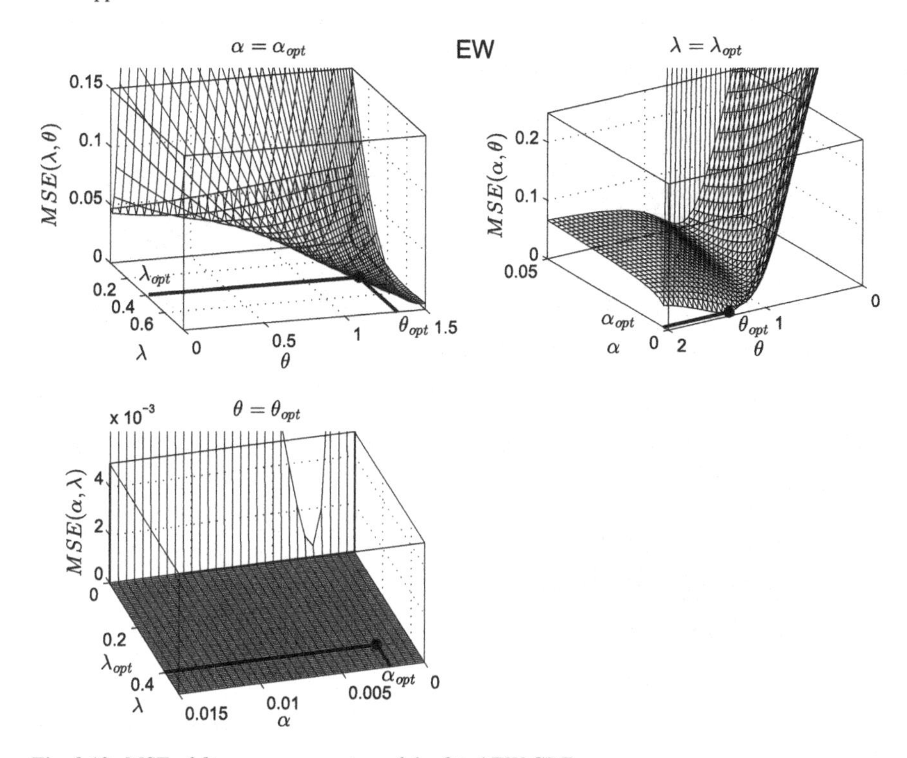

Fig. 3.12 MSE of fit versus parameters of the fitted EW CDF

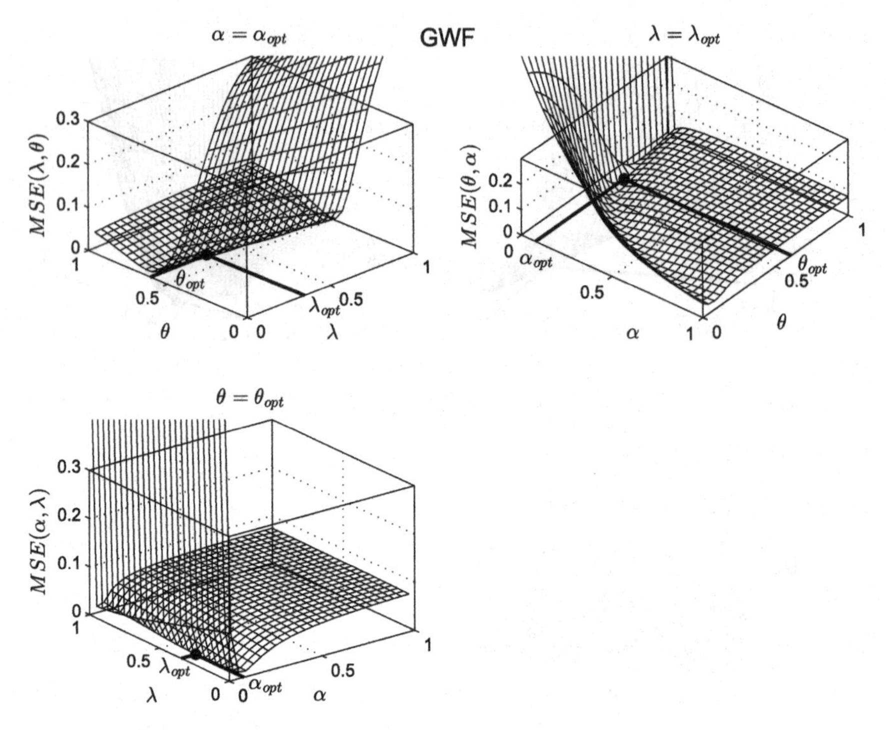

Fig. 3.13 MSE of fit versus parameters of the fitted GWF CDF

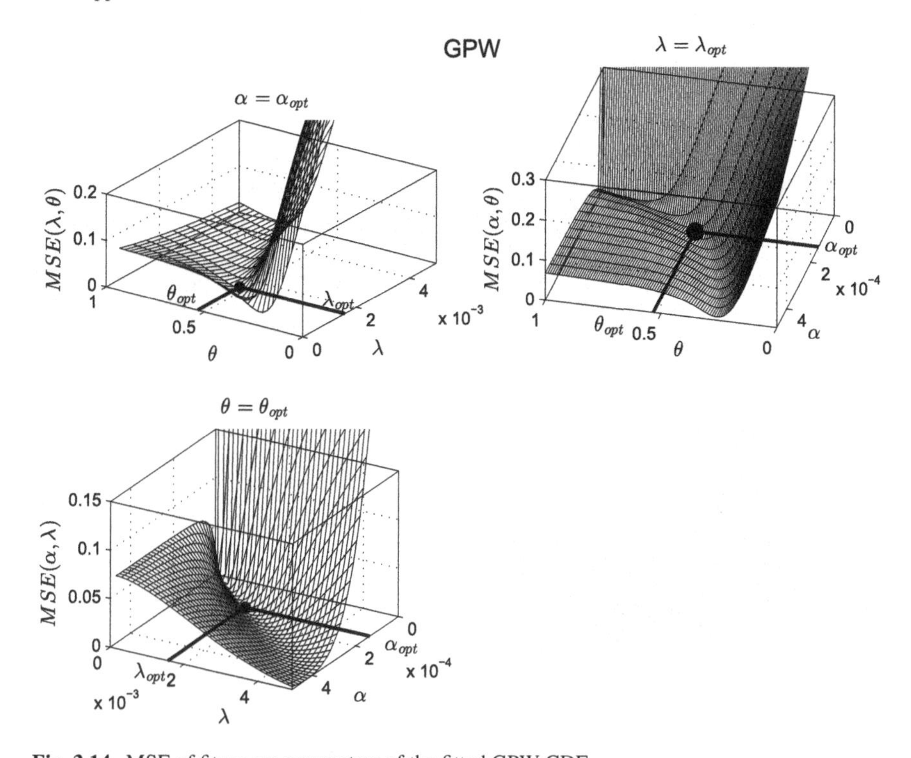

Fig. 3.14 MSE of fit versus parameters of the fitted GPW CDF

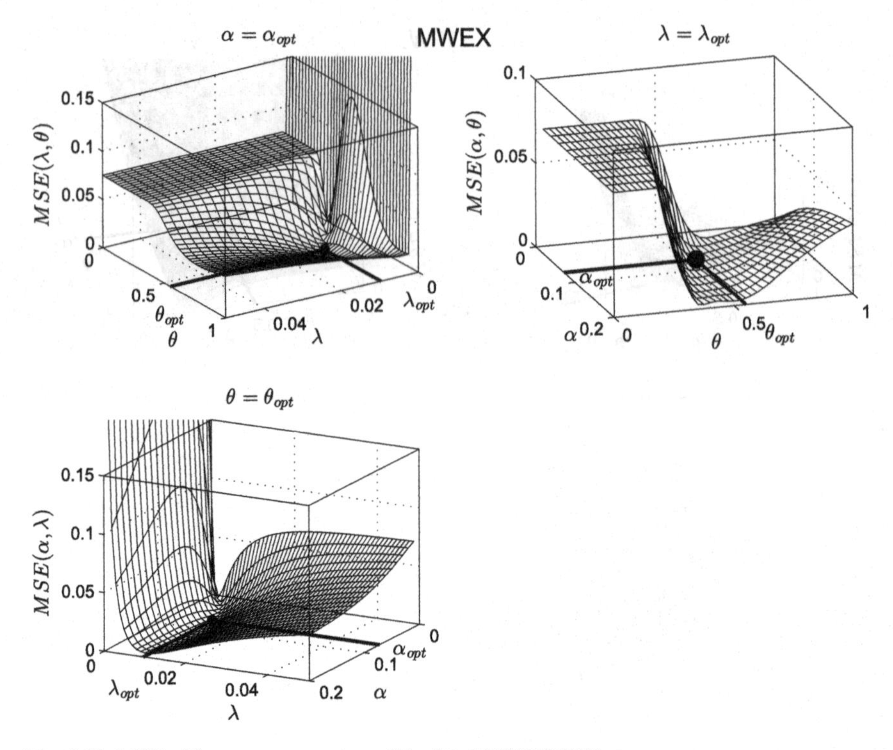

Fig. 3.15 MSE of fit versus parameters of the fitted MWEX CDF

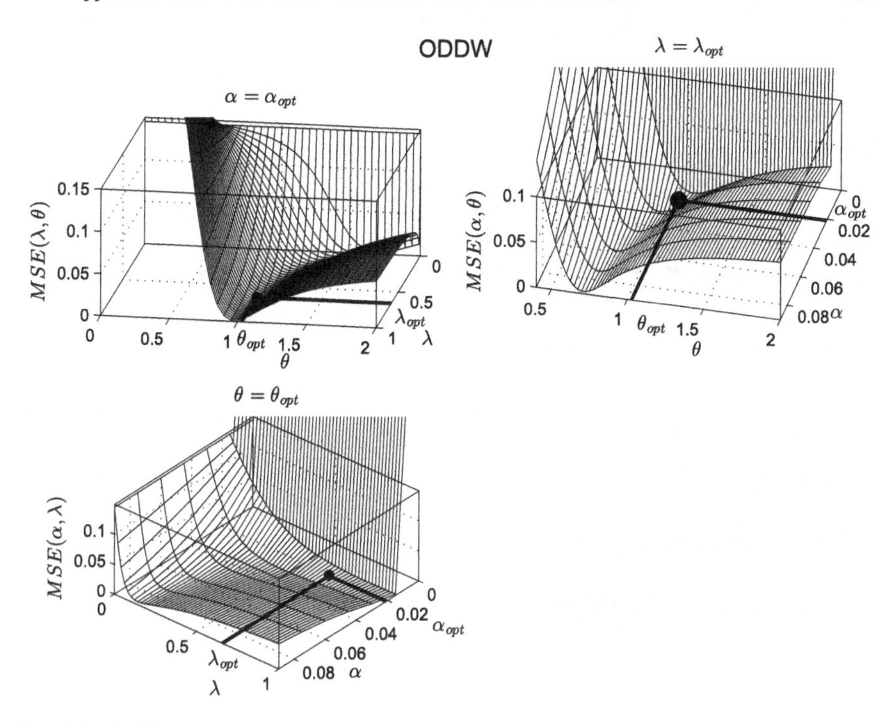

Fig. 3.16 MSE of fit versus parameters of the fitted ODDW CDF

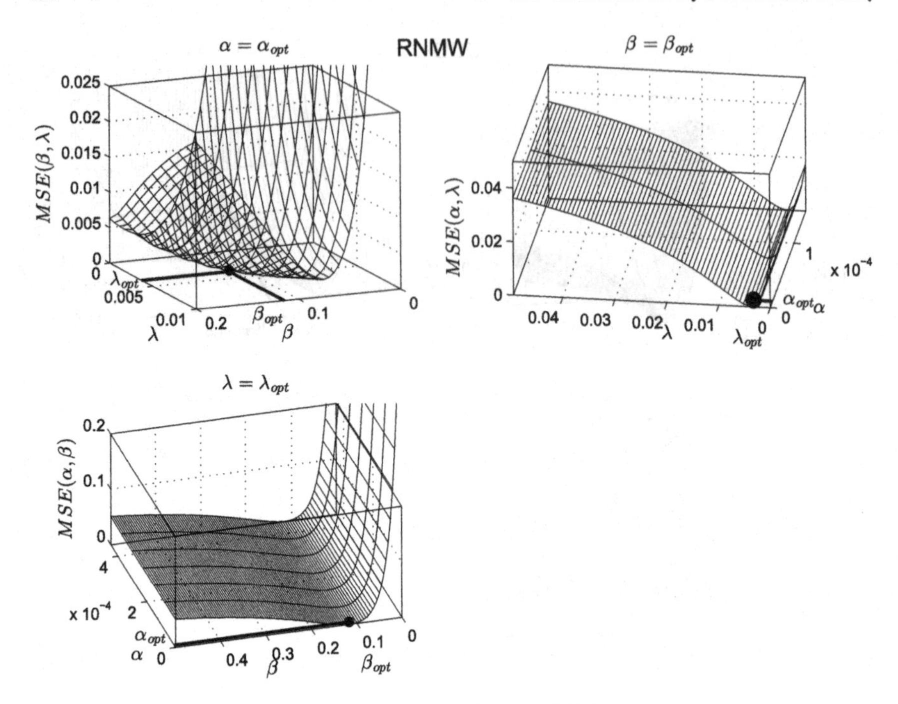

Fig. 3.17 MSE of fit versus parameters of the fitted RNMW CDF

References

1. J. Dombi, T. Jónás, Z.E. Tóth, The epsilon probability distribution and its application in reliability theory. Acta Polytech. Hung. **15**(1), 197–216 (2018a)
2. J. Dombi, T. Jónás, Z.E. Tóth, G. Árva, The omega probability distribution and its applications in reliability theory. Qual. Reliab. Eng. Int. **35**(2), 600–626 (2019). https://doi.org/10.1002/qre.2425
3. J. Dombi, T. Jónás, Approximations to the normal probability distribution function using operators of continuous-valued logic. Acta Cybern. **23**(3), 829–852 (2018a)
4. J. Dombi, T. Jónás, Kappa regression: an alternative to logistic regression. Int. J. Uncertain. Fuzziness Knowl.-Based Syst. **28**(02), 237–267 (2020a). https://doi.org/10.1142/S0218488520500105
5. J. Dombi, On a certain type of unary operators, in *2012 IEEE International Conference on Fuzzy Systems* (2012), pp. 1–7. https://doi.org/10.1109/FUZZ-IEEE.2012.6251349
6. A. Papoulis, S.U. Pillai, *Probability, Random Variables, and Stochastic Processes*, 4th edn. (McGraw-Hill Higher Education, Upper Saddle River, 2002)
7. W. Weibull, A statistical distribution function of wide applicability. J. Appl. Mech. **18**, 293–297 (1951)
8. I.E. Okorie, S. Nadarajah, On the omega probability distribution. Quality Reliab. Eng. Int. **0**(0) (2019). https://doi.org/10.1002/qre.2462
9. T. Csendes, L. Pál, O.H. Sendin, J.R. Bangha, The global optimization method revisited. Optim. Lett. **2**(4), 445–454 (2008)
10. T. Csendes, Nonlinear parameter estimation by global opitmization - efficiency and reliability. Acta Cybern. **8**(4), 361–372 (1988)

11. D. Murthy, M. Xie, R. Jiang, *Weibull Models*, Wiley Series in Probability and Statistics (Wiley, New York, 2004)
12. J. Kies, N.R.L. (U.S.), N.P.S. (U.S.), *The Strength of Glass*. NRL report. Naval Research Laboratory (1958)
13. K.K. Phani, A new modified Weibull distribution function. J. Am. Ceram. Soc. **70**(8), C–182–C–184 (1987). https://doi.org/10.1111/j.1151-2916.1987.tb05719.x
14. G.S. Mudholkar, D.K. Srivastava, G.D. Kollia, A generalization of the Weibull distribution with application to the analysis of survival data. J. Am. Stat. Assoc. **91**(436), 1575–1583 (1996). https://doi.org/10.1080/01621459.1996.10476725
15. M. Xie, C. Lai, Reliability analysis using an additive Weibull model with bathtub-shaped failure rate function. Reliab. Eng. Syst. Saf. **52**(1), 87–93 (1996). https://doi.org/10.1016/0951-8320(95)00149-2
16. T. Zhang, M. Xie, Failure data analysis with extended Weibull distribution. Commun. Stat. - Simul. Comput. **36**(3), 579–592 (2007). https://doi.org/10.1080/03610910701236081
17. M.E. Ghitany, E.K. Al-Hussaini, R.A. Al-Jarallah, Marshall-Olkin extended Weibull distribution and its application to censored data. J. Appl. Stat. **32**(10), 1025–1034 (2005). https://doi.org/10.1080/02664760500165008
18. M. Ghitany, D. Al-Mutairi, N. Balakrishnan, L. Al-Enezi, Power lindley distribution and associated inference. Comput. Stat. Data Anal. **64**, 20–33 (2013). https://doi.org/10.1016/j.csda.2013.02.026
19. C.D. Lai, M. Xie, D.N.P. Murthy, A modified Weibull distribution. IEEE Trans. Reliab. **52**(1), 33–37 (2003). https://doi.org/10.1109/TR.2002.805788
20. A.M. Sarhan, J. Apaloo, Exponentiated modified Weibull extension distribution. Reliab. Eng. Syst. Saf. **112**, 137–144 (2013). https://doi.org/10.1016/j.ress.2012.10.013
21. G.O. Silva, E.M.M. Ortega, G.M. Cordeiro, The beta modified Weibull distribution. Lifetime Data Anal. **16**(3), 409–430 (2010). https://doi.org/10.1007/s10985-010-9161-1
22. M. Bebbington, C. Lai, R. Zitikis, A flexible Weibull extension. Reliab. Eng. Syst. Saf. **92**(6), 719–726 (2007). https://doi.org/10.1016/j.ress.2006.03.004
23. G.M. Cordeiro, E.M. Ortega, S. Nadarajah, The Kumaraswamy Weibull distribution with application to failure data. J. Frankl. Inst. **347**(8), 1399–1429 (2010). https://doi.org/10.1016/j.jfranklin.2010.06.010
24. S.J. Almalki, S. Nadarajah, Modifications of the Weibull distribution: a review. Reliab. Eng. Syst. Saf. **124**, 32–55 (2014)
25. S. SC, *Reliability, Life Testing and the Prediction of Service Lives: For Engineers and Scientists* (Springer Science & Business Media, Berlin, 2007)
26. G.S. Mudholkar, D.K. Srivastava, Exponentiated Weibull family for analyzing bathtub failure-rate data. IEEE Trans. Reliab. **42**(2), 299–302 (1993). ISSN 0018-9529. https://doi.org/10.1109/24.229504
27. M. GS, K. GD, Generalized Weibull family: a structural analysis. Commun. Stat. - Theory Methods **23**(4), 1149–1171 (1994). https://doi.org/10.1080/03610929408831309
28. M. Nikulin, F. Haghighi, A chi-squared test for the generalized power Weibull family for the head-and-neck cancer censored data. J. Math. Sci. **133**(3), 1333–1341 (2006)
29. M. Xie, Y. Tang, T.N. Goh, A modified Weibull extension with bathtub-shaped failure rate function. Reliab. Eng. Syst. Saf. **76**(3), 279–285 (2002). https://doi.org/10.1016/S0951-8320(02)00022-4
30. K. Cooray, Generalization of the Weibull distribution: the odd Weibull family. Stat. Modell. **6**(3), 265–277 (2006). https://doi.org/10.1191/1471082X06st116oa
31. S. Almalki, A reduced new modified Weibull distribution (2013)
32. T. Yuge, M. Maruyama, S. Yanagi, Reliability of a k-out-of-n system with common-cause failures using multivariate exponential distribution. Proc. Comput. Sci. **96**, 968–976 (2016). https://doi.org/10.1016/j.procs.2016.08.101
33. L. Wang, Y. Shi, Reliability analysis of a class of exponential distribution under record values. J. Comput. Appl. Math. **239**, 367–379 (2013). https://doi.org/10.1016/j.cam.2012.09.004

34. N. Johnson, S. Kotz, N. Balakrishnan, *Continuous Univariate Distributions*, Wiley Series in Probability and Mathematical Statistics: Applied Probability and Statistics, vol. 2 (Wiley, New York, 1995)
35. N. Balakrishnan, *Handbook of the Logistic Distribution*, Statistics: A Series of Textbooks and Monographs (Taylor & Francis, Milton Park, 2013)
36. S.M. Mousavi, S.P. Horton, C.A. Langston, B. Samei, Seismic features and automatic discrimination of deep and shallow induced-microearthquakes using neural network and logistic regression. Geophys. J. Int. **207**(1), 29–46 (2016). https://doi.org/10.1093/gji/ggw258
37. R.-Z. Liu, Z.-R. Zhao, C.S. Ng, Statistical modelling for thoracic surgery using a nomogram based on logistic regression. J. Thorac. Dis. **8**(8), E731–6 (2016). ISSN 2072-1439. https://doi.org/10.21037/jtd.2016.07.91
38. R.J. Klement, J. Belderbos, I. Grills, M. Werner-Wasik, A. Hope, M. Giuliani, H. Ye, H. Sonke, J.-J. Peulen, M. Guckenberger, Prediction of early death in patients with early-stage nsclc-can we select patients without a potential benefit of sbrt as a curative treatment approach? J. Thorac. Oncol.: Off. Publ. Int. Assoc. Study of Lung Cancer **11**(7), 1132–1139 (2016). https://doi.org/10.1016/j.jtho.2016.03.016
39. U.I. Ahmed, L. Ying, M.K. Bashir, M. Abid, E. Elahi, M. Iqbal, Access to output market by small farmers: the case of Punjab, Pakistan. J. Animal Plant Sci. **26**(3), 787–793 (2016)
40. H. Lu, H. Liyan, Z. Hongwei, Combined model of empirical study for credit risk management, in *2010 2nd IEEE International Conference on Information and Financial Engineering (ICIFE)*, pp. 189–192. IEEE (2010)
41. G. Casella, R. Berger, *Statistical Inference* (Duxbury Resource Center, 2001)
42. S. Chatterjee, A.S. Hadi, *Regression Analysis by Example* (Wiley, New York, 2015)
43. F. Harrell, *Regression Modeling Strategies. With Applications to Linear Models, Logistic and Ordinal Regression, and Survival Analysis*, 2nd edn. (Springer International Publishing, Berlin, 2015)
44. D.W. Hosmer Jr., S. Lemeshow, R.X. Sturdivant, *Applied Logistic Regression*, vol. 398 (Wiley, New York, 2013)
45. I. Matic, R. Radoicic, D. Stefanica, A sharp Pólya-based approximation to the normal cdf. SSRN (2016). https://doi.org/10.2139/ssrn.2842681
46. A. Soranzo, E. Epure, Very simply explicitly invertible approximations of normal cumulative and normal quantile function. Appl. Math. Sci. **8**(87), 4323–4341 (2014)
47. R. Yerukala, N.K. Boiroju, Approximating to the cumulative normal function and its inverse. Int. J. Sci. Eng. Res. **6**(4), 515–518 (2015)
48. K.D. Tocher, *The Art of Simulation* (English University Press, London, 1963)
49. F.S. Hillier, G.J. Lieberman, *Introduction to Operations Research*, 7th edn. (McGraw-Hill, New York, 2001)
50. S. Bowling, M. Khasawneh, S. Kaewkuekool, B. Cho, A logistic approximation to the cumulative normal distribution. J. Ind. Eng. Manag. **2**(1), 114–127 (2009)
51. G. Pólya, Remarks on computing the probability integral in one and two dimensions, in *Proceedings of the 1st Berkeley Symposium on Mathematical Statistics and Probability* (University of California Press, Berkeley, California, 1949) pp. 63–78
52. J.-T. Lin, A simpler logistic approximation to the normal tail probability and its inverse. Appl. Stat. **39**, 255–257 (1990)
53. K.M. Aludaat, M.T. Alodat, A note on approximating the normal distribution function. Appl. Math. Sci. **2**(9), 425–429 (2008)
54. O. Eidous, S. Al-Salman, One-term approximation for normal distribution function. Math. Stat. **4**(1), 15–18 (2016)

Chapter 4
A Fuzzy Arithmetic-Based Time Series Model

From a practical perspective, a time series is a collection of values of certain events or tasks which are obtained with respect to time [1]. Predicting future values of time series that represent sales, stock price, product demand, level of inventory, revenue and profit are primary issues in the field of forecasting. Due to the dynamic nature of time series data, which is highly non-stationary and uncertain, the decision making process may become quite wearisome [2]. In order to overcome this problem, various time series forecasting approaches have been proposed in the literature. The most notable issues in time series forecasting are related to the fitting of models, quantity of data, improving the efficiency of models, influencing factors, statistical applicability of results, data transformation and consequences of prediction. All the above-mentioned issues suggest that intelligent forecasting techniques are required here [3].

In modeling real-life time series, the applications of fuzzy sets [4] or fuzzy numbers [5] seem to be an appropriate approach [6]. One of the main advantages of fuzzy logic based approaches is that they do not need to satisfy strict assumptions such as having a large sample, linear model, stationary and normal distribution, which are formulated as elementary requirements in traditional time series approaches [7]. The fuzzy time series approach, which is based on the fuzzy set theory introduced by Zadeh [6], was first proposed by Song and Chissom [8–10] for developing a model to describe the uncertainty and imprecise knowledge involved in time series data. They initially used the fuzzy sets concept to represent or manage all these uncertainties, and referred to this concept as fuzzy time series (FTS). From 1994 onwards, numerous fuzzy time series models have been developed both for short-term [11, 12] and long-term forecasting [13]. The FTS methods provide generally remarkable forecasting performance [7] and these methods can be applied in cases where just small sets of data are available. Recently, several soft computing techniques

J. Dombi and T. Jónás, *Advances in the Theory of Probabilistic and Fuzzy Data Scientific Methods with Applications*, Studies in Computational Intelligence 814, https://doi.org/10.1007/978-3-030-51949-0_4

including fuzzy clustering, artificial neural networks and genetic algorithms have been employed to improve further the forecasting performance of FTS methods [3, 14]. These soft computing techniques can provide significant solution for domain specific problems. The combination of these techniques leads to the development of a new architecture, which is more advantageous providing robust, cost effective and approximate solution in comparison to conventional techniques [1, 3]. The main objective of the application of hybrid techniques is to find solutions to nonlinear problems which are difficult to model using traditional approaches [7, 15–20].

Fuzzy time series methods generally consist of three stages; namely, fuzzification of crisp observations, determination of fuzzy relations, and defuzzification, all playing an important role in the performance of the applied model. Current studies in the field aim to contribute to these three steps by providing some improvements to the fuzzy time series methods in determining the lengths of intervals, fuzzification process and defuzzification techniques. The fuzzification step [7] includes the decomposition of the universe of discourse by determining intervals [6, 17, 21–28]. In respect of defining fuzzy relations, various methods focusing on this step have been proposed [6, 17, 29–34]. Concerning the defuzzification step, the centroid methods [21, 30] and the adaptive expectation methods [25, 35] are two widely used approaches.

The fuzzy time series modeling technique proposed in this chapter (see also [36]) is based on a fuzzy inference method in which the fuzzy output is a so-called pliant number. This fuzzy inference system is called the pliant arithmetic-based fuzzy inference system (PAFIS). The defuzzification method of this systems has two notable advantages. Firstly, it does not require any numerical integration to generate the crisp output; and secondly, it runs in a constant time. Afterwards, we discuss how the pliant arithmetic-based fuzzy time series (PAFTS) model can be established by utilizing the PAFIS method. The greatest advantages of our PAFTS model lies in its simplicity and easy-to-use characteristics. Once the fuzzy rule consequents are obtained, the time series modeling and forecasting become very simple. In order to evaluate the forecasting performance of the PAFTS method, 17 time series were analyzed including the Australian Beer Consumption (ABC) from 1956 to 1994, the Istanbul Stock Exchange Market (BIST100) Index for the time period between 2009 and 2013, and the Taiwan Stock Exchange Capitalization Weighted Stock Index (TAIEX) for the time period between 1999 and 2004. The forecasting results of our method were compared with those of the Winter's multiplicative exponential smoothing (WMES), seasonal autoregressive integrated moving average (SARIMA), feed-forward artificial neural network (FFANN), adaptive neuro-fuzzy inference system (ANFIS), modified adaptive network-based fuzzy inference system (MANFIS) [37], autoregressive adaptive network fuzzy inference system (AR-ANFIS) [38] methods, and with those of Chen [30], Chen and Chang [39], Chen and Chen [40] and Chen et al. [41]. Based on our empirical results, the PAFTS method may be viewed as novel viable time series modeling technique.

The rest of this chapter is structured as follows. Section 4.1 discusses the theoretical basis of pliant numbers. In Sect. 4.2, the pliant arithmetic-based fuzzy inference system is introduced in details. In Sect. 4.3, the proposed method is utilized for fuzzy time series modeling purposes. Here, the pliant arithmetic-based fuzzy time series model is applied to some well-known time series and the forecasting performance of the proposed method is compared to those of some recent advanced time series modeling methods. Lastly, in Sect. 4.4 we summarize our main conclusions. This chapter is connected with the following publications of ours: [36, 42–45].

4.1 The Concept of Pliant Numbers

In our pliant arithmetic-based fuzzy inference system, which we will discuss in Sect. 4.2, the consequents of fuzzy rules are special quasi fuzzy numbers. These quasi fuzzy numbers are the so-called pliant numbers. Here, the theoretical basis of these fuzzy numbers will be discussed. Following Dombi's pliant inequality model [46], the following notations will be used. The indexes l and r stand for 'left' and 'right', respectively, and they will be used to denote left and right hand side components of fuzzy numbers. As we will discuss consequents of fuzzy rules, we will use the notation y for the crisp output variable.

4.1.1 Pliant Numbers

Definition 4.1 The sigmoid function $\sigma_a^{(\lambda)}(y)$ with the parameters a and λ is given by

$$\sigma_a^{(\lambda)}(y) = \frac{1}{1 + \mathrm{e}^{-\lambda(y-a)}}, \tag{4.1}$$

where $y, a, \lambda \in \mathbb{R}$ and λ is nonzero.

Notice that the sigmoid function in Eq. (4.1) is a linearly transformed version of the logistic probability distribution function given in Eq. (3.31). Also, the logistic regression function in Eq. (3.43) can be written in the same form as the sigmoid function in Eq. (4.1).

The main properties of the sigmoid function $\sigma_a^{(\lambda)}(y)$ are as follows. The function is monotone increasing, if λ is positive, and it is monotone decreasing, if λ is negative.

$$\lim_{y \to +\infty} \sigma_a^{(\lambda)}(y) = \begin{cases} 1, & \text{if } \lambda > 0 \\ 0, & \text{if } \lambda < 0 \end{cases}$$

$$\lim_{y \to -\infty} \sigma_a^{(\lambda)}(y) = \begin{cases} 1, & \text{if } \lambda < 0 \\ 0, & \text{if } \lambda > 0 \end{cases}$$

$\sigma_a^{(\lambda)}(a) = 0.5$, the slope of $\sigma_a^{(\lambda)}(y)$ in the $(a, 0.5)$ point is determined by λ as

$$\frac{d\sigma_a^{(\lambda)}(a)}{dy} = \frac{\lambda}{4}.$$

Definition 4.2 The truth values of the inequalities $a_l < y$ and $a_r > y$ are given by the sigmoid functions

$$\{a_l <_{(\lambda_l)} y\} = \sigma_{a_l}^{(\lambda_l)}(y) = \frac{1}{1 + e^{-\lambda_l(y-a_l)}}$$

and

$$\{a_r >_{(\lambda_r)} y\} = \sigma_{a_r}^{(\lambda_r)}(y) = \frac{1}{1 + e^{-\lambda_r(y-a_r)}},$$

respectively, where $a_l, \lambda_l, a_r, \lambda_r \in \mathbb{R}$, $\lambda_l > 0$, $\lambda_r < 0$, $y \in \mathbb{R}$.

In the previous two definition, the parameters λ_l and λ_r determine the sharpness of the soft inequalities $\{a_l <_{(\lambda_l)} y\}$ and $\{a_r >_{(\lambda_r)} y\}$, respectively. Notice that λ_l is positive and λ_r is negative. The next theorem demonstrates a key property of $\{a_l <_{(\lambda_l)} y\}$-like soft inequalities that are given by sigmoid functions.

Theorem 4.1 *For any* $\alpha \in (0, 1)$, *if* $w_1, w_2, \ldots, w_n \geq 0$, $\lambda_{l,1}, \lambda_{l,2}, \ldots, \lambda_{l,n} > 0$ *and* $\{a_{l,1} <_{(\lambda_{l,1})} y\} = \{a_{l,2} <_{(\lambda_{l,2})}\} = \cdots = \{a_{l,n} <_{(\lambda_{l,n})} y\} = \alpha$, *then* $\left\{a_l <_{(\lambda_l)} y\right\} = \alpha$, *where*

$$a_l = \sum_{i=1}^{n} w_i a_{l,i}, \qquad \frac{1}{\lambda_l} = \sum_{i=1}^{n} \frac{w_i}{\lambda_{l,i}}.$$

Proof Utilizing the definition of the soft inequality $\{a_{l,i} <_{(\lambda_{l,i})} y\}$ and the condition that $\{a_{l,i} <_{(\lambda_{l,i})} y\} = \alpha$, we get

$$\{a_{l,i} <_{(\lambda_{l,i})} y\} = \sigma_{a_{l,i}}^{(\lambda_{l,i})}(y) = \frac{1}{1 + e^{-\lambda_{l,i}(y-a_{l,i})}} = \alpha, \tag{4.2}$$

$i = 1, 2, \ldots, n$. Here, we will use the notation $\sigma\left(y; a_{l,i}, \lambda_{l,i}\right)$ for $\sigma_{a_{l,i}}^{(\lambda_{l,i})}(y)$; that is,

$$\sigma\left(y; a_{l,i}, \lambda_{l,i}\right) = \sigma_{a_{l,i}}^{(\lambda_{l,i})}(y).$$

Since the function $\sigma\left(y; a_{l,i}, \lambda_{l,i}\right)$ is strictly monotonic, it is invertible. Its inverse function $\sigma^{-1}\left(y; a_{l,i}, \lambda_{l,i}\right)$ is

$$\sigma^{-1}\left(y; a_{l,i}, \lambda_{l,i}\right) = -\ln\left(\frac{1-y}{y}\right)\frac{1}{\lambda_{l,i}} + a_{l,i}.$$

Utilizing the function $\sigma^{-1}\left(y; a_{l,i}, \lambda_{l,i}\right)$, the $y = y_i$ that satisfies the condition in Eq. (4.2) is

$$y_i = \sigma^{-1}\left(\alpha; a_{l,i}, \lambda_{l,i}\right) = -\ln\left(\frac{1-\alpha}{\alpha}\right)\frac{1}{\lambda_{l,i}} + a_{l,i}. \tag{4.3}$$

Hence,

$$\sum_{i=1}^{n} w_i y_i = \sum_{i=1}^{n} w_i \sigma^{-1}\left(\alpha; a_{l,i}, \lambda_{l,i}\right) =$$

$$= -\ln\left(\frac{1-\alpha}{\alpha}\right)\sum_{i=1}^{n}\frac{w_i}{\lambda_{l,i}} + \sum_{i=1}^{n} w_i a_{l,i} = -\ln\left(\frac{1-\alpha}{\alpha}\right)\frac{1}{\lambda_l} + a_l,$$

where

$$a_l = \sum_{i=1}^{n} w_i a_{l,i}, \qquad \frac{1}{\lambda_l} = \sum_{i=1}^{n}\frac{w_i}{\lambda_{l,i}}.$$

Next,

$$-\ln\left(\frac{1-\alpha}{\alpha}\right)\frac{1}{\lambda_l} + a_l$$

has the same form as the right hand side of Eq. (4.3), so

$$-\ln\left(\frac{1-\alpha}{\alpha}\right)\frac{1}{\lambda_l} + a_l = \sigma^{-1}\left(\alpha; a_l, \lambda_l\right).$$

Exploiting this result, $y = \sigma^{-1}(\alpha; a_l, \lambda_l)$ is the point where $\sigma(y; a_l, \lambda_l) = \alpha$; that is $\left\{a_l <_{(\lambda_l)} y\right\} = \alpha$. □

This result means that if the 'a is less than y' type soft inequalities are represented by sigmoid functions with positive λ parameters, then the weighted sum of these inequalities can also be represented by a sigmoid function. Figure 4.1 shows the weighted sum of two $\{a_l <_{(\lambda_l)} y\}$-like soft inequalities when the sum of the non-negative weights is equal to one. Similar to Theorem 4.1, the following theorem on the weighted sum of 'a is greater than y' type soft inequalities can also be proven.

Theorem 4.2 *For any* $\alpha \in (0, 1)$, *if* $w_1, w_2, \ldots, w_n \geq 0$, $\lambda_{r,1}, \lambda_{r,2}, \ldots, \lambda_{r,n} < 0$ *and* $\{a_{r,1} >_{(\lambda_{r,1})} y\} = \{a_{r,2} >_{(\lambda_{r,2})} y\} = \cdots = \{a_{r,n} >_{(\lambda_{r,n})} y\} = \alpha$, *then* $\left\{a_r >_{(\lambda_r)} y\right\} = \alpha$, *where*

$$a_r = \sum_{i=1}^{n} w_i a_{r,i}, \qquad \frac{1}{\lambda_r} = \sum_{i=1}^{n}\frac{w_i}{\lambda_{r,i}}.$$

In our method, a pliant number will be composed of an increasing left hand side and a decreasing right hand side sigmoid function. Theorems 4.1 and 4.2 allow us to separately sum the corresponding sides of pliant numbers, and finally combine them into one pliant number, which is the result of fuzzy arithmetic addition of the original pliant numbers. Since the sigmoid function does not take the value of 1, we will introduce its following alternative variant, which allows us to set its value arbitrarily close to 1 at a given point.

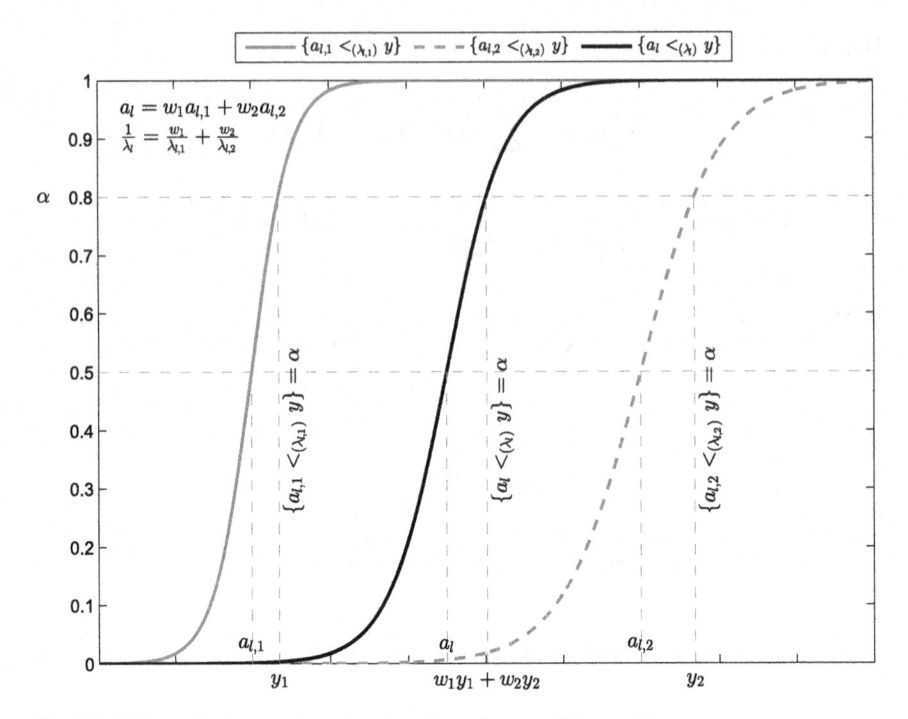

Fig. 4.1 The weighted sum of two 'a is less than y' type soft inequalities

Let ε be a small positive number and let $\varepsilon < 0.5$, for example $\varepsilon = 0.001$. Furthermore, let $a_l < y_0 < a_r$, $\lambda_l > 0$, $\lambda_r < 0$. If we wish $\sigma_{a_l}^{(\lambda_l)}(y)$ and $\sigma_{a_r}^{(\lambda_r)}(y)$ to satisfy the criterion

$$\sigma_{a_l}^{(\lambda_l)}(y_0) = \sigma_{a_r}^{(\lambda_r)}(y_0) = 1 - \varepsilon,$$

we need to set λ_l and λ_r as follows:

$$\lambda_l = \frac{\ln\left(\frac{1-\varepsilon}{\varepsilon}\right)}{y_0 - a_l},$$

$$\lambda_r = \frac{\ln\left(\frac{1-\varepsilon}{\varepsilon}\right)}{y_0 - a_r}.$$

In this case,

$$\sigma_{a_l}^{(\lambda_l)}(y) = \frac{1}{1 + \left(\frac{\varepsilon}{1-\varepsilon}\right)^{\frac{y-a_l}{y_0-a_l}}} = \sigma(y; a_l, y_0, \varepsilon), \tag{4.4}$$

$$\sigma_{a_r}^{(\lambda_r)}(y) = \frac{1}{1 + \left(\frac{\varepsilon}{1-\varepsilon}\right)^{\frac{y-a_r}{y_0-a_r}}} = \sigma(y; y_0, a_r, \varepsilon). \tag{4.5}$$

From here on, we will use the notations $\sigma(y; a_l, y_0, \varepsilon)$ and $\sigma(y; y_0, a_r, \varepsilon)$ for the functions $\sigma_{a_l}^{(\lambda_l)}(y)$ and $\sigma_{a_r}^{(\lambda_r)}(y)$, when these are given by Eqs. (4.4) and (4.5), respectively. Utilizing $\sigma(y; a_l, y_0, \varepsilon)$ and $\sigma(y; y_0, a_r, \varepsilon)$, the fuzzy membership function $\mu(y, y_0, a_l, a_r, \varepsilon)$ is defined as follows.

Definition 4.3 The fuzzy membership function $\mu(y; a_l, y_0, a_r, \varepsilon)$ is given by

$$\mu(y; a_l, y_0, a_r, \varepsilon) = \begin{cases} \sigma(y; a_l, y_0, \varepsilon), & \text{if } y < y_0 \\ \sigma(y; y_0, a_r, \varepsilon), & \text{if } y \geq y_0, \end{cases}$$

where $a_l < y_0 < a_r, 0 < \varepsilon < 0.5$.

Since

$$\mu(y_0; a_l, y_0, a_r, \varepsilon) = 1 - \varepsilon,$$

$$\lim_{y \to -\infty} \mu(y; a_l, y_0, a_r, \varepsilon) = 0, \qquad \lim_{y \to +\infty} \mu(y; a_l, y_0, a_r, \varepsilon) = 0,$$

the membership function $\mu(y; a_l, y_0, a_r, \varepsilon)$ represents quite well the soft equation $\{y = y_0\}$, if ε is close to zero ($\varepsilon > 0$). It means that $\mu(y; a_l, y_0, a_r, \varepsilon)$ may be viewed as the membership function of the quasi fuzzy number 'y is equal to y_0'. (Note that function $\mu(y; a_l, y_0, a_r, \varepsilon)$ could theoretically be the membership function of the fuzzy number 'y is equal to y_0', if $\mu(y; a_l, y_0, a_r, \varepsilon)$ took the value of 1 at y_0.) A great advantage of the membership function $\mu(y; a_l, y_0, a_r, \varepsilon)$ is that its both sides are given by sigmoid functions; and so, Theorems 4.1 and 4.2 can be utilized to sum quasi fuzzy numbers that are given by $\mu(y; a_l, y_0, a_r, \varepsilon)$-like membership functions. We should also add that the function $\mu(y; a_l, y_0, a_r, \varepsilon)$ is not differentiable at y_0; its graph is 'peaked' there (see Fig. 4.2). Utilizing the functions $\sigma(y; a_l, y_0, \varepsilon)$ and $\sigma(y; y_0, a_r, \varepsilon)$, the soft inequalities $\{a_l <_{(\lambda_l)} y\}$ and $\{a_r >_{(\lambda_r)} y\}$ can be written in the following forms:

$$\{a_l <_{(\lambda_l)} y\} = \{a_l <_{(y_0,\varepsilon)} y\} = \sigma(y; a_l, y_0, \varepsilon)$$

$$\{a_r >_{(\lambda_r)} y\} = \{a_r >_{(y_0,\varepsilon)} y\} = \sigma(y; y_0, a_r, \varepsilon),$$

where $a_l < y_0 < a_r, 0 < \varepsilon < 0.5$. Furthermore, Theorems 4.1 and 4.2 can be turned into Proposition 4.1 and Proposition 4.2, respectively.

Proposition 4.1 *For any* $\alpha \in (0, 1)$, *if* $w_1, w_2, \ldots, w_n \geq 0$, $\lambda_{l,1}, \lambda_{l,2}, \ldots, \lambda_{l,n} > 0$ *and* $\{a_{l,1} <_{(y_{0,1},\varepsilon)} y\} = \{a_{l,2} <_{(y_{0,2},\varepsilon)} y\} = \cdots = \{a_{l,n} <_{(y_{0,n},\varepsilon)} y\} = \alpha$, *then* $\left\{a_l <_{(y_0,\varepsilon)} y\right\} = \alpha$, *where*

$$a_l = \sum_{i=1}^{n} w_i a_{l,i}, \qquad y_0 = \sum_{i=1}^{n} w_i y_{0,i}.$$

Proposition 4.2 *For any $\alpha \in (0, 1)$, if $w_1, w_2, \ldots, w_n \geq 0$, $\lambda_{r,1}, \lambda_{r,2}, \ldots, \lambda_{r,n} < 0$ and $\{a_{r,1} >_{(y_{0,1},\varepsilon)} y\} = \{a_{r,2} >_{(y_{0,2},\varepsilon)} y\} = \cdots = \{a_{r,n} >_{(y_{0,n},\varepsilon)} y\} = \alpha$, then $\left\{a_r >_{(y_0,\varepsilon)} y\right\} = \alpha$, where*

$$a_r = \sum_{i=1}^{n} w_i a_{r,i}, \qquad y_0 = \sum_{i=1}^{n} w_i y_{0,i}.$$

Furthermore, Dombi's conjunction operator [47] will be utilized to derive the soft inequality $\{a_l <_{(y_0,\varepsilon)} y <_{(y_0,\varepsilon)} a_r\}$ as the intersection of the soft inequalities $\{a_l <_{(y_0,\varepsilon)} y\}$ and $\{a_r >_{(y_0,\varepsilon)} y\}$.

Definition 4.4 (Dombi's t-norm) Let A_1 and A_2 be fuzzy sets defined over the crisp universe $Y \subseteq \mathbb{R}$ given by the membership functions $\mu_{A_1}(y)$ and $\mu_{A_2}(y)$, respectively. The Dombi intersection of A_1 and A_2 is given by the membership function $\mu_{A_1 \cap A_2}(y)$:

$$\mu_{A_1 \cap A_2}(y) = \mu_{A_1}(y) *_{(D)} \mu_{A_2}(y) =$$
$$= \frac{1}{1 + \left(\left(\frac{1-\mu_{A_1}(y)}{\mu_{A_1}(y)} \right)^{\alpha} + \left(\frac{1-\mu_{A_2}(y)}{\mu_{A_2}(y)} \right)^{\alpha} \right)^{1/\alpha}},$$

where $\alpha \in \mathbb{R}$, $\alpha > 0$, and $*_{(D)}$ denotes the Dombi intersection operator.

Applying the Dombi intersection with $\alpha = 1$ to $\sigma(y; a_l, y_0, \varepsilon)$ and $\sigma(y; y_0, a_r, \varepsilon)$, the soft inequality $\{a_l <_{(a_l,y_0,\varepsilon)} y <_{(y_0,a_r,\varepsilon)} a_r\}$ has the following $\mu_p(y; y_0, a_l, a_r, \varepsilon)$ fuzzy membership function:

$$\{a_l <_{(a_l,y_0,\varepsilon)} y <_{(y_0,a_r,\varepsilon)} a_r\} = \mu_p(y; a_l, y_0, a_r, \varepsilon) =$$
$$= \sigma(y; a_l, y_0, \varepsilon) *_{(D)} \sigma(y; y_0, a_r, \varepsilon) = \frac{1}{1 + \left(\frac{\varepsilon}{1-\varepsilon}\right)^{\frac{y-a_l}{y_0-a_l}} + \left(\frac{\varepsilon}{1-\varepsilon}\right)^{\frac{y-a_r}{y_0-a_r}}}.$$

Figure 4.2 shows the plots of two sigmoid functions and the plots of the membership functions $\mu(y; a_l, y_0, a_r, \varepsilon)$ and $\mu_p(y; a_l, y_0, a_r, \varepsilon)$ composed of the two sigmoid functions.

Figure 4.3 suggests that the function $\mu_p(y; a_l, y_0, a_r, \varepsilon)$ is very similar to the function $\mu(y; a_l, y_0, a_r, \varepsilon)$ and so $\mu_p(y; a_l, y_0, a_r, \varepsilon)$ may be viewed as the membership function of a quasi fuzzy number.

Proposition 4.3 *If $0 < \varepsilon < 0.5$, then*

$$\max_{y \in \mathbb{R}} \left| \mu_p(y; a_l, y_0, a_r, \varepsilon) - \mu(y; a_l, y_0, a_r, \varepsilon) \right| \leq \varepsilon \frac{1-\varepsilon}{1+\varepsilon},$$

where $a_l < y_0 < a_r$.

Proof Notice that for any $y \in \mathbb{R}$, $\mu_p(y; a_l, y_0, a_r, \varepsilon) < \mu(y; a_l, y_0, a_r, \varepsilon)$ and so

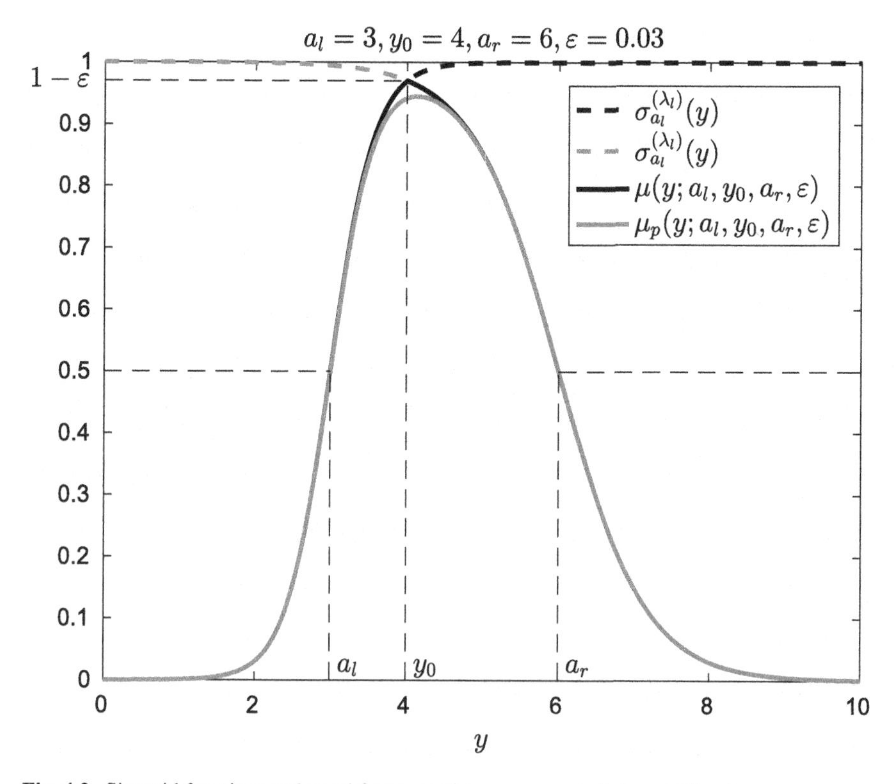

Fig. 4.2 Sigmoid functions and quasi fuzzy numbers composed of sigmoid functions

$$\left|\mu_p(y; a_l, y_0, a_r, \varepsilon) - \mu(y; a_l, y_0, a_r, \varepsilon)\right| = \mu(y; a_l, y_0, a_r, \varepsilon) - \mu_p(y; a_l, y_0, a_r, \varepsilon).$$

Let the function $d\colon \mathbb{R} \to \mathbb{R}$ be given by

$$d(y) = \mu(y; a_l, y_0, a_r, \varepsilon) - \mu_p(y; a_l, y_0, a_r, \varepsilon).$$

Let $0 < \varepsilon < 0.5$. It can be shown that if $y \geq y_0$, then the function d is monotonously decreasing; that is, $d(y) \leq d(y_0)$ for any $y \geq y_0$. Noting the fact that

$$d(y_0) = \varepsilon \frac{1-\varepsilon}{1+\varepsilon}, \tag{4.6}$$

we have

$$d(y) \leq \varepsilon \frac{1-\varepsilon}{1+\varepsilon}$$

for $y \geq y_0$.

Next, if $y \leq y_0$, then the function d is monotonously increasing. That is, $d(y) \leq d(y_0)$ for any $y \leq y_0$. In this case, by noting Eq. (4.6), we have

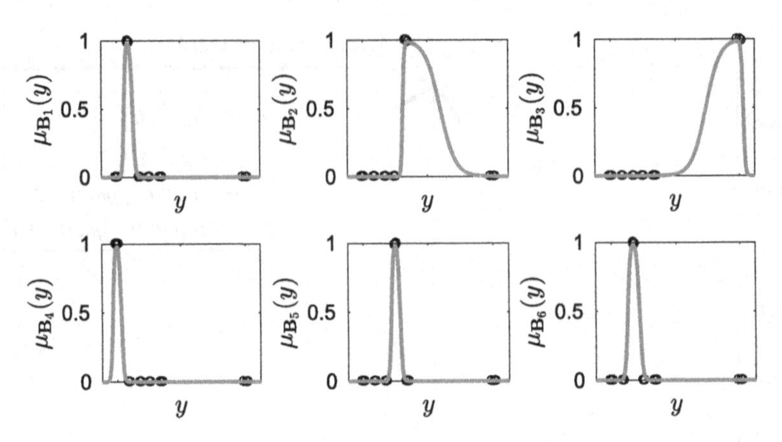

Fig. 4.3 Examples of pliant numbers as fuzzy rule consequents

$$d(y) \le d(y_0) = \varepsilon \frac{1-\varepsilon}{1+\varepsilon}.$$

□

Corollary 4.1 *For any* $y \in \mathbb{R}$

$$\lim_{\varepsilon \to 0} \left| \mu_p(y; a_l, y_0, a_r, \varepsilon) - \mu(y; a_l, y_0, a_r, \varepsilon) \right| = 0,$$

where $a_l < y_0 < a_r$, $\varepsilon > 0$.

Proof The corollary follows from Proposition 4.3. □

Proposition 4.3 and Corollary 4.1 indicate that if ε is small ($\varepsilon > 0$), then the membership function $\mu_p(y; a_l, y_0, a_r, \varepsilon)$ approximates quite well the membership function $\mu(y; a_l, y_0, a_r, \varepsilon)$. Therefore, in practice, the membership function $\mu_p(y; a_l, y_0, a_r, \varepsilon)$ may be viewed as the membership function of a quasi fuzzy number that we call the pliant number. It should be added here that unlike the function $\mu(y; a_l, y_0, a_r, \varepsilon)$, the function $\mu_p(y; a_l, y_0, a_r, \varepsilon)$ is differentiable at any y.

Definition 4.5 (*Pliant number*) The pliant number 'y is equal to y_0' is given by the membership function

$$\mu_p(y; a_l, y_0, a_r, \varepsilon) = \frac{1}{1 + \left(\frac{\varepsilon}{1-\varepsilon}\right)^{\frac{y-a_l}{y_0-a_l}} + \left(\frac{\varepsilon}{1-\varepsilon}\right)^{\frac{y-a_r}{y_0-a_r}}},$$

where $a_l < y_0 < a_r$, $0 < \varepsilon < 0.5$.

Later in our pliant arithmetic-based fuzzy inference system, the following lemma and corollary will be utilized.

Lemma 4.1 *The membership function $\mu_p(y; a_l, y_0, a_r, \varepsilon)$ is maximal, if*

$$y = y_0 + \frac{\ln\left(\frac{a_r - y_0}{y_0 - a_l}\right)}{\ln\left(\frac{1-\varepsilon}{\varepsilon}\right)} \frac{(y_0 - a_l)(a_r - y_0)}{a_r - a_l},$$

where $a_l < y_0 < a_r$, $0 < \varepsilon < 0.5$.

Proof The function $\mu_p(y; a_l, y_0, a_r, \varepsilon)$ is maximal, if the function

$$g(y) = \left(\frac{\varepsilon}{1-\varepsilon}\right)^{\frac{y-a_l}{y_0-a_l}} + \left(\frac{\varepsilon}{1-\varepsilon}\right)^{\frac{y-a_r}{y_0-a_r}}$$

is minimal. Since the function $g(y)$ is the sum of two strictly convex differentiable functions, there is only one $\hat{y}$ for which $g(y)$ is minimal. $\hat{y}$ can be obtained by solving the equation

$$\frac{\mathrm{d}g(y)}{\mathrm{d}y} = \ln\left(\frac{\varepsilon}{1-\varepsilon}\right)\left(\frac{\left(\frac{\varepsilon}{1-\varepsilon}\right)^{\frac{y-a_l}{y_0-a_l}}}{y_0 - a_l} - \frac{\left(\frac{\varepsilon}{1-\varepsilon}\right)^{\frac{y-a_r}{y_0-a_r}}}{a_r - y_0}\right) = 0.$$

From this equation

$$\hat{y} = y_0 + \frac{\ln\left(\frac{a_r - y_0}{y_0 - a_l}\right)}{\ln\left(\frac{1-\varepsilon}{\varepsilon}\right)} \frac{(y_0 - a_l)(a_r - y_0)}{a_r - a_l}. \tag{4.7}$$

□

This result can be interpreted as the pliant number 'y is equal to y_0' is the quasi fuzzy number 'y is equal to $\hat{y}$'. Moreover, $\hat{y}$ is close to y_0, if ε has a sufficiently small positive value. The next corollary demonstrates this property of the pliant number 'y is equal to y_0'.

Corollary 4.2 *If $\varepsilon \to 0$, then the membership function $\mu_p(y; a_l, y_0, a_r, \varepsilon)$ is maximal at y_0, where $a_l < y_0 < a_r$, $\varepsilon > 0$.*

Proof Based on Lemma 4.1, $\mu_p(y; a_l, y_0, a_r, \varepsilon)$ is maximal at $\hat{y}$ that is given by Eq. (4.7). Next, if $\varepsilon \to 0$, then

$$\ln\left(\frac{1-\varepsilon}{\varepsilon}\right) \to \infty,$$

and so $\hat{y} \to y_0$. □

Notice that if the pliant number given by the membership function $\mu_p(y; a_l, y_0, a_r, \varepsilon)$ is symmetric, then $y_0 - a_l = a_r - y_0$, from which we get that $y_0 = (a_l + a_r)/2$. In this case, Eq. (4.7) tells us that the maximum of $\mu_p(y; a_l, y_0, a_r, \varepsilon)$ is at $\hat{y} =$

y_0. This result is in line with the expectation that if the curve of $\mu_p(y; a_l, y_0, a_r, \varepsilon)$ is symmetric, then the place of its maximum is the midpoint of the (a_l, a_r) interval.

4.2 Pliant Arithmetic-Based Fuzzy Inference Systems

Our goal is to construct a fuzzy inference system (FIS) that can model the time series $X_1, X_2, \ldots, X_n$. Let $X_j, X_{j+1}, \ldots, X_{j+r-1}$ be an r-period-long sub-time series and X_{j+r} its one period long continuation in the time series $X_1, X_2, \ldots, X_n$ $(j = 1, 2, \ldots, n-r)$. Here, r is the number of historical periods that is used for generating one-period-ahead forecast, $(r \geq 1, r+1 \leq n)$. Next, let $s = n - r$ denote the number of r-period-long sub-time series in $X_1, X_2, \ldots, X_n$. Here, we will introduce the pliant arithmetic-based fuzzy inference system (PAFIS), which will be then utilized for time series modeling in Sect. 4.3. In the subsections below, we will discuss how the PAFIS inference system is constructed.

4.2.1 *Data Preparation*

For an r-period-long sub-time series $X_j, X_{j+1}, \ldots, X_{j+r-1}$ and its X_{j+r} continuation, let the vector $\mathbf{x}_j = (x_{j,1}, x_{j,2}, \ldots, x_{j,r})$ and the scalar y_j be defined as follows $(j = 1, 2, \ldots, s)$.

1) If $X_j, X_{j+1}, \ldots, X_{j+r-1}$ are not all equal, then

$$x_{j,p} = \frac{X_{j+p-1} - \min\limits_{q=1,\ldots,r}(X_{j+q-1})}{\max\limits_{q=1,\ldots,r}(X_{j+q-1}) - \min\limits_{q=1,\ldots,r}(X_{j+q-1})}$$

$$y_j = \frac{X_{j+r} - \min\limits_{q=1,\ldots,r}(X_{j+q-1})}{\max\limits_{q=1,\ldots,r}(X_{j+q-1}) - \min\limits_{q=1,\ldots,r}(X_{j+q-1})}$$

$p = 1, 2, \ldots, r$.

2) If $X_j, X_{j+1}, \ldots, X_{j+r-1}$ are all equal and nonzero, say have the value of a $(a > 0)$, then $\mathbf{x}_j = (1, 1, \ldots, 1)$, and $y_j = X_{j+r}/a$.
3) If $X_j, X_{j+1}, \ldots, X_{j+r-1}$ are all zeros, then $\mathbf{x}_j = (0, 0, \ldots, 0)$, and $y_j = X_{j+r}$.

Owing to this transformation, each component of vector $\mathbf{x}_j$ is normalized to the interval [0, 1]. The $(\mathbf{x}_j, y_j)$ pairs represent the time series $X_1, X_2, \ldots, X_n$ so that the normalized vector $\mathbf{x}_j$ is followed by y_j, and our goal is to build a FIS that can adequately map the relation between $\mathbf{x}_j$ and y_j $(j = 1, 2, \ldots, s)$. So, our aim is to build a fuzzy inference system that has an r-dimensional vector $\mathbf{x}$ as input, and a scalar output y. The system approximates the $y = f(\mathbf{x})$ relation based on the sample $(\mathbf{x}_j, y_j)$, where $\mathbf{x}_j$, y_j are the corresponding observations on $\mathbf{x}$ and y, respectively. In order to identify typical patterns in the time series, the $\mathbf{x}_j$ vectors are clustered using the fuzzy c-means clustering algorithm [48].

4.2.2 Forming Fuzzy Rules

Let ϕ be the exponent used in the fuzzy c-means clustering algorithm ($\phi \in \mathbb{R}$, $1 < \phi < \infty$) that determines the fuzziness of clusters, and let $\mathbf{C}_1, \mathbf{C}_2, \ldots, \mathbf{C}_m$ be the clusters formed with the cluster centroids $\mathbf{c}_1, \mathbf{c}_2, \ldots, \mathbf{c}_m$, respectively ($1 \leq m \leq s$). (The typical value of ϕ used in practice is 2.)

The $\mathbf{C}_1, \mathbf{C}_2, \ldots, \mathbf{C}_m$ fuzzy clusters represent fuzzy partitions in the input space. From a time series perspective, the r-period-long normalized sub-time series are classified into m clusters, and the cluster centroids represent r-period-long typical normalized patterns. Based on this, the following fuzzy rules can be formed.

$$
\begin{array}{llll}
\text{Rule } 1: & \text{if } \mathbf{x} \text{ is in } \mathbf{C}_1 & \text{then} & \mathbf{B}_1 \\
\text{Rule } 2: & \text{if } \mathbf{x} \text{ is in } \mathbf{C}_2 & \text{then} & \mathbf{B}_2 \\
\vdots & \vdots & \vdots & \vdots \\
\text{Rule } m: & \text{if } \mathbf{x} \text{ is in } \mathbf{C}_m & \text{then} & \mathbf{B}_m,
\end{array}
$$

where $\mathbf{x} \in \mathbb{R}^r$ is an input vector normalized according to the method described in Sect. 4.2.1, and $\mathbf{B}_1, \mathbf{B}_2, \ldots, \mathbf{B}_m$ are fuzzy sets defined over set $\mathbf{Y}$ which is the domain of crisp system outputs. For any r-dimensional normalized input vector $\mathbf{x}$, its membership value $\mu_i(\mathbf{x})$ in cluster $\mathbf{C}_i$ can be calculated as follows:

$$
\mu_i(\mathbf{x}) = \left(\frac{1}{\|\mathbf{x} - \mathbf{c}_i\|_2^2} \right)^{\frac{1}{\phi-1}} \frac{1}{\sum\limits_{v=1}^{m} \left(\frac{1}{\|\mathbf{x}-\mathbf{c}_v\|_2^2} \right)^{\frac{1}{\phi-1}}}. \tag{4.8}
$$

The value of $\mu_i(\mathbf{x})$ may be viewed as the activation level of Rule i for the input $\mathbf{x}$. As a result of fuzzy c-means clustering, the rule antecedents have been identified.

In our method, each $\mathbf{B}_i$ consequent is treated as a pliant number with the membership function $\mu_{\mathbf{B}_i}(y)$. In order to make the fuzzy inference system complete, the consequent of each rule needs to be identified; that is, the membership function of each pliant number $\mathbf{B}_i$ needs to be supplied.

4.2.3 Membership Functions of Fuzzy Rule Consequents

Let the $\mathbf{Y}$ domain of the crisp outputs of our FIS be

$$\mathbf{Y} = [y_l - \Delta, y_h + \Delta],$$

where $y_l = \min_{j=1,...,s}(y_j)$, $y_h = \max_{j=1,...,s}(y_j)$, $\Delta = c(y_h - y_l)$, $c > 0$, $c \in \mathbb{R}$. In our implementation, $c = 0.1$.

Now, simple heuristics are introduced to identify the parameter of each membership function $\mu_{\mathbf{B}_i}(y)$ $(i = 1, 2, \ldots, m)$. Here, we have the sample of pairs $(\mathbf{x}_j, y_j)$ in which $\mathbf{x}_j$ is the jth input and y_j is the corresponding output $(j = 1, 2, \ldots, s)$. As mentioned before, the activation level $\mu_i(\mathbf{x}_j)$ of the ith fuzzy rule antecedent for input $\mathbf{x}_j$ may be calculated according to Eq. (4.8); that is, we have the pairs $(y_j, \mu_i(\mathbf{x}_j))$. Next, we have the membership function $\mu_{\mathbf{B}_i}(y) = \mu_{p,i}(y; a_{l,i}, y_{0,i}, a_{r,i}, \varepsilon)$, where $a_{l,i} < y_{0,i} < a_{r,i}, 0 < \varepsilon < 0.5$ $(i = 1, 2, \ldots, m)$. In our implementation, $\varepsilon = 0.001$. The values of parameters $a_{l,i}, y_{0,i}, a_{r,i}$ are obtained by minimizing the quantity

$$\delta_i = \sum_{j=1}^{s} \left(\mu_{p,i}(y; a_{l,i}, y_{0,i}, a_{r,i}, \varepsilon) - \mu_i(\mathbf{x}_j)\right)^2.$$

Here, δ_i can be minimized by using the interior point algorithm [49]. In this algorithm, we initialize the parameters $y_{0,i}$, $a_{l,i}$ and $a_{r,i}$ as follows:

$$y_{0,i} = \frac{\sum_{j=1}^{s} \mu_i(\mathbf{x}_j) y_j}{\sum_{j=1}^{s} \mu_i(\mathbf{x}_j)} \tag{4.9}$$

$$a_{l,i} = \frac{y_l + y_{0,i}}{2} \tag{4.10}$$

$$a_{r,i} = \frac{y_h + y_{0,i}}{2}. \tag{4.11}$$

Figure 4.3 shows some examples of pliant numbers as fuzzy rule consequents obtained by applying the above-described method.

4.2.4 Aggregation of Fuzzy Outputs

The $\mu_i(\mathbf{x})$ quantity given by Eq. (4.8) measures how much the input vector $\mathbf{x}$ activates the ith fuzzy rule. In other words, $\mu_i(\mathbf{x})$ may be interpreted as the level of applicability of rule i to input $\mathbf{x}$. Based on this, the normalized weight of rule i, w_i, for input $\mathbf{x}$ is

$$w_i = \frac{\mu_i(\mathbf{x})}{\sum_{k=1}^{m} \mu_k(\mathbf{x})}, \text{ if } \sum_{k=1}^{m} \mu_k(\mathbf{x}) \neq 0. \tag{4.12}$$

Notice that nothing is inferred, if $\sum_{k=1}^{m} \mu_k(\mathbf{x}) = 0$.

When we build a pliant arithmetic-based fuzzy inference system, the fuzzy rule consequents are pliant numbers; that is, $\mu_{\mathbf{B}_i}(y) = \mu_{p,i}(y; a_{l,i}, y_{0,i}, a_{r,i}, \varepsilon)$, where $a_{l,i} < y_{0,i} < a_{r,i}$, $0 < \varepsilon < 0.5$ $(i = 1, 2, \ldots, m)$. In this case, based on Proposition 4.3 and Corollary 4.1, the left-hand side and the right-hand side of each membership function $\mu_{p,i}(y; a_{l,i}, y_{0,i}, a_{r,i}, \varepsilon)$ is approximately an increasing and a decreasing sigmoid function, respectively. Utilizing this property of each $\mu_{p,i}(y; a_{l,i}, y_{0,i}, a_{r,i}, \varepsilon)$ and the Propositions 4.1 and 4.2, the aggregate fuzzy output $\mu_{\mathbf{B}}(y)$ of our PAFIS is the pliant number with the membership function $\mu_p(y; a_l, y_0, a_r, \varepsilon)$, where

$$a_l = \sum_{i=1}^{m} w_i a_{l,i}, \qquad y_0 = \sum_{i=1}^{m} w_i y_{0,i}, \qquad a_r = \sum_{i=1}^{m} w_i a_{r,i}.$$

It should be emphasized here that there is no implication applied in our inference methods. The fuzzy outputs of the PAFIS method are pliant numbers obtained via fuzzy arithmetic operations; namely, via weighted summation of pliant numbers.

4.2.5 *Defuzzification*

As the aim is to use our pliant arithmetic-based fuzzy inference systems for time series modeling, a defuzzification method, which transforms the fuzzy output **B** into a crisp $\hat{y}$, needs to be identified. Here, we will use a maximum defuzzification method; that is, we will represent the fuzzy output **B**, which is given by the membership function $\mu_{\mathbf{B}}(y)$, by the crisp $\hat{y}$ for which

$$\mu_{\mathbf{B}}(\hat{y}) = \max_{y \in \mathbf{Y}} \mu_{\mathbf{B}}(y).$$

When we use the pliant arithmetic-based inference, the fuzzy output is a pliant number; that is, $\mu_{\mathbf{B}}(y) = \mu_p(y; a_l, y_0, a_r, \varepsilon)$, where $a_l < y_0 < a_r$, $0 < \varepsilon < 0.5$. In this case, based on Lemma 4.1, the place $\hat{y}$, where $\mu_p(y; a_l, y_0, a_r, \varepsilon)$ is maximal can be directly identified from its parameters:

$$\hat{y} = y_0 + \frac{\ln\left(\frac{a_r - y_0}{y_0 - a_l}\right)}{\ln\left(\frac{1-\varepsilon}{\varepsilon}\right)} \frac{(y_0 - a_l)(a_r - y_0)}{a_r - a_l}.$$

Recall that based on Corollary 4.2, if ε is close to zero, then $\hat{y} \approx y_0$.

The introduced defuzzification method has two notable advantages. Firstly, it does not require any numerical integration to generate the crisp output; and secondly, it runs in a constant time.

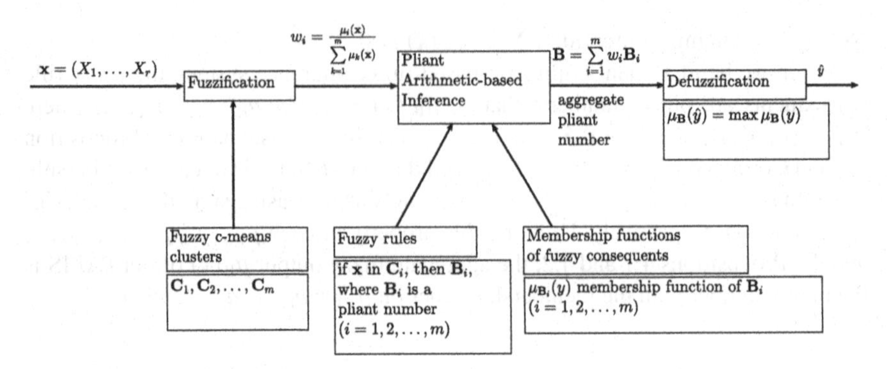

Fig. 4.4 The block-scheme of PAFIS method

4.2.6 *The Inference Mechanism*

The inference method of the PAFIS system is summarized in the block-scheme shown in Fig. 4.4. Here, we will give a brief description of the pliant arithmetic-based inference procedure.

The input to the inference system is an r-dimensional vector $\mathbf{x} = (X_1, X_2, \ldots, X_r)$. The fuzzification method utilizes the fuzzy c-means clusters that were generated for a training data set of r-dimensional vectors. Here, the fuzzification results in the membership value $\mu_i(\mathbf{x})$ of vector $\mathbf{x}$ in cluster $\mathbf{C}_i$, $i = 1, 2, \ldots, m$. The value of $\mu_i(\mathbf{x})$, which can be computed as in Eq. (4.8), measures how much the input vector $\mathbf{x}$ activates the ith rule of the fuzzy rule-base. Once the value of $\mu_i(\mathbf{x})$ is computed for all $i = 1, 2, \ldots, m$, the normalized weight w_i of the ith rule for the input vector $\mathbf{x}$ is calculated as in Eq. (4.12). Recall that the consequent $\mathbf{B}_i$ of the ith fuzzy rule is a pliant number, the membership function of which can be obtained as described in Sect. 4.2.3. Next, the pliant arithmetic inference produces the aggregate fuzzy output $\mathbf{B} = \sum_{i=1}^{m} w_i \mathbf{B}_i$, which is a pliant number as well based on Sect. 4.2.4. Lastly, the defuzzification step converts the aggregate fuzzy output $\mathbf{B}$ into the crisp output $\hat{y}$ by applying the defuzzification method described in Sect. 4.2.5.

It is worth mentioning that the fuzzy output is obtained via weighted averaging of pliant numbers; that is, the inference is based on a fuzzy arithmetic operation rather than on logical implications.

4.3 Forecasting with the Pliant Arithmetic-Based Fuzzy Time Series Model

Let $X_1, X_2, \ldots, X_r$ be an r-period-long known time series and X_{r+1} be its one-period-long unknown continuation, $r \geq 1$. Furthermore, let $\hat{X}_{r+1}$ denote the forecast for X_{r+1} generated based on the $X_1, X_2, \ldots, X_r$ values using a fuzzy inference system constructed based on the method that was presented in Sect. 4.2. Here, $\hat{X}_{r+1}$ can be generated as follows. Depending on the $X_1, X_2, \ldots, X_r$ values, the normalized input vector $\mathbf{x} = (x_1, x_2, \ldots, x_r)$ to PAFIS is created by applying one of the following cases.

(1) If $X_1, X_2, \ldots, X_r$ are not all equal, then

$$x_p = \frac{X_p - \min\limits_{q=1,\ldots,r}(X_q)}{\max\limits_{q=1,\ldots,r}(X_q) - \min\limits_{q=1,\ldots,r}(X_q)}$$

$p = 1, 2, \ldots, r.$

(2) If $X_1, X_2, \ldots, X_r$ are equal and nonzero, say have the value of a ($a > 0$), then $\mathbf{x} = (1, 1, \ldots, 1)$.

(3) If $X_1, X_2, \ldots, X_r$ are all zeros, then $\mathbf{x} = (0, 0, \ldots, 0)$.

Let $\hat{y}$ be the PAFIS output for the input vector $\mathbf{x}$. Depending on how vector $\mathbf{x}$ was created from the time series $X_1, X_2, \ldots, X_r$ based on the cases above, the $\hat{X}_{r+1}$ forecast is computed according to one of the following denormalizations.

(1) If $x_1, x_2, \ldots, x_r$ are not all equal, then

$$\hat{X}_{r+1} = \hat{y}\left(\max\limits_{q=1,\ldots,r}(X_q) - \min\limits_{q=1,\ldots,r}(X_q)\right) + \min\limits_{q=1,\ldots,r}(X_q).$$

(2) If $\mathbf{x} = (1, 1, \ldots, 1)$, then $\hat{X}_{r+1} = a\hat{y}$. (For definition of a, see Sect. 4.2.1.)

(3) If $\mathbf{x} = (0, 0, \ldots, 0)$, then $\hat{X}_{r+1} = \hat{y}$.

If we apply this forecasting method for X_{j+r} continuation of each r-period-long sub-time series $X_j, X_{j+1}, \ldots, X_{j+r-1}$ in the time series $X_1, X_2, \ldots, X_n$, then $\hat{X}_{r+1}, \hat{X}_{r+2}, \ldots, \hat{X}_n$ are the predicted (simulated) values of $X_{r+1}, X_{r+2}, \ldots, X_n$, respectively. Namely, the $\hat{X}_{r+1}, \hat{X}_{r+2}, \ldots, \hat{X}_n$ values model the $X_{r+1}, X_{r+2}, \ldots, X_n$ ones ($j = 1, 2, \ldots, n - r$).

4.3.1 Demonstrative Examples

The pliant arithmetic-based fuzzy time series model (PAFTS) was implemented in MatLab R2017b. Following the recent paper of Sarıca et al. [38], in which they proposed the so-called autoregressive adaptive network fuzzy inference system (AR-ANFIS), 17 time series were analyzed in our study to evaluate the forecasting performance of the PAFTS method. The first time series contains the quarterly data of Australian Beer Consumption (ABC) from 1956 to 1994. The next 10 time series contain daily data of the Istanbul Stock Exchange Market (BIST100) Index for the time period between 2009 and 2013, two time series for each year with different

Table 4.1 Properties of the analyzed time series and the parameters of PAFTS models

Identifier (ID)	Series name	Length of series	# of training data	# of test data	# of lag	# of clusters
1	ABC	148	132	16	4	16
2	BIST100:2009	103	96	7	15	15
3	BIST100:2009	103	88	15	5	15
4	BIST100:2010	104	97	7	14	15
5	BIST100:2010	104	89	15	3	13
6	BIST100:2011	106	99	7	4	9
7	BIST100:2011	106	91	15	4	10
8	BIST100:2012	106	99	7	4	12
9	BIST100:2012	106	91	15	4	12
10	BIST100:2013	106	99	7	3	9
11	BIST100:2013	106	91	15	3	7
12	TAIEX:1999	266	221	45	3	14
13	TAIEX:2000	271	224	47	5	11
14	TAIEX:2001	244	201	43	4	7
15	TAIEX:2002	248	205	43	4	10
16	TAIEX:2003	249	206	43	6	18
17	TAIEX:2004	250	205	45	7	15

numbers of training data. The last 6 time series include daily data of the Taiwan Stock Exchange Capitalization Weighted Stock Index (TAIEX) for the time period between 1999 and 2004. The properties of the examined time series and the parameters of the PAFTS method for each of the time series are summarized in Table 4.1. Analyzing these time series allows us to compare the forecasting performance of the PAFTS methods with that of some recent methods discussed by Sarıca et al. [38].

In the analyses below, the Calinski-Harabasz [50] index was used to find the quasi optimal number of clusters (number of fuzzy rules) for the PAFTS method. The examined methods were compared by utilizing the root-mean-square error (RMSE) and the mean absolute percentage error (MAPE) performance indicators that are given by Eqs. (4.13) and (4.14), respectively. That is,

$$RMSE = \sqrt{\frac{1}{n}\sum_{i=1}^{n}\left(X_i - \hat{X}_i\right)^2} \tag{4.13}$$

$$MAPE = \frac{1}{n}\sum_{i=1}^{n}\left|\frac{X_i - \hat{X}_i}{X_i}\right|, \tag{4.14}$$

where $X_1, X_2, \ldots, X_n$ and $\hat{X}_1, \hat{X}_2, \ldots, \hat{X}_n$ are the actual values and the forecast values, respectively.

4.3.1.1 Australian Beer Consumption Time Series

The ABC time series contains 148 data out of which the first 132 data were used for training, while the rest 16 data were utilized for testing purposes; that is, forecasts for 16 periods were computed. For this time series, the best PAFTS results were obtained with 16 clusters and a number of lag of 4 (number of historical periods $r = 4$). The forecasting results of the PAFTS method were compared with those of the AR–ANFIS, Winter's multiplicative exponential smoothing (WMES), seasonal autoregressive integrated moving average (SARIMA), feed-forward artificial neural network (FFANN), adaptive neuro-fuzzy inference system (ANFIS) and modified adaptive network based fuzzy inference system (MANFIS) [37] methods. These forecasting results are shown in Table 4.2. Note that results of the AR–ANFIS method are quoted from [38] and the results of the other above-mentioned methods, except for the PAFTS, are referenced from the paper of Eğrioğlu et al. [37].

The underlined numbers in Table 4.2 show the best RMSE and MAPE results. The test data and the PAFTS forecast results for the ABC time series are shown in

Table 4.2 Simulation results for ABC time series

Test data	WMES	SARIMA	FFANN	ANFIS	MANFIS	AR–ANFIS	PAFTS
430.50	453.91	452.72	453.8838	446.7141	451.456	445.231	452.31
600.00	575.22	578.29	557.8151	553.7286	575.632	589.867	593.37
464.50	502.32	487.7	497.5159	482.0749	494.066	486.806	506.33
423.60	444.73	446.28	437.393	434.1992	436.565	441.308	430.18
437.00	459.66	456.77	449.0035	438.5516	444.696	450.353	437.79
574.00	582.48	583.51	569.0025	559.006	575.423	571.947	577.34
443.00	508.64	492.13	471.0758	472.5198	481.273	469.621	488.48
410.00	450.31	450.36	424.3307	427.5736	414.437	428.886	423.28
420.00	465.4	461.01	448.8667	445.0127	430.307	419.428	426.47
532.00	589.74	588.96	560.0436	562.9364	565.178	570.378	552.5
432.00	514.96	496.77	447.0135	459.1439	452.052	440.589	460.91
420.00	455.89	454.64	408.6362	416.1582	392.138	400.060	409.74
411.00	471.15	465.46	428.1073	431.7013	419.331	413.925	429.36
512.00	597.00	594.71	537.6988	544.9831	536.878	549.318	516.5
449.00	521.28	501.67	438.433	444.3101	446.324	441.676	452.08
382.00	461.46	459.17	420.5827	426.01	406.637	413.590	412.83
RMSE	53.3295	47.0367	24.1052	25.0500	21.3728	20.5791	21.3460
MAPE	0.1072	0.0949	0.0476	0.0467	0.0400	0.0376	0.0373

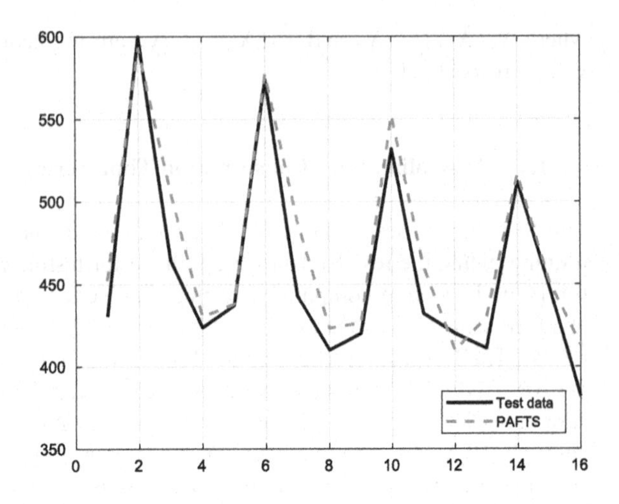

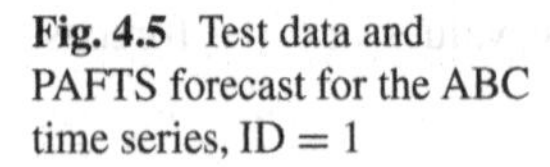
Fig. 4.5 Test data and PAFTS forecast for the ABC time series, ID = 1

Fig. 4.5. It can be seen from Table 4.2 that the AR–ANFIS method gave the best RMSE value (20.5791) and the PAFTS method resulted in the second best RMSE value (21.3460) compared with the other studied methods. This table also tells us that PAFTS produced the lowest MAPE value (0.0373), while the second lowest MAPE value (0.0376) was obtained by AR–ANFIS. Based on these points, we may conclude that for the ABC time series, the PAFTS and AR–ANFIS methods both produce similarly good forecasts.

4.3.1.2 Istanbul Stock Exchange Market (BIST100) Index

The forecasting performance of the PAFTS and QPAFTS methods were examined using 10 time series of daily data of the Istanbul Stock Exchange Market (BIST100) Index for the time period between 2009 and 2013. For each of these five years, two analyses were conducted with 7 and 15 forecast periods, respectively. Therefore, there are two time series for each year of the BITS100 data in Table 4.1 (ID = 2,...,11) with different numbers of training and test data. The performance of the PAFTS method was compared with those of the ARIMA, ANFIS, MANFIS and AR–ANFIS methods. The results of the ARIMA method are quoted from Baş et al. [7], the ANFIS, MANFIS and AR–ANFIS results were taken from the paper of Sarıca et al. [38]. The RMSE and MAPE values of the compared methods are summarized in Table 4.3, where the best RMSE and MAPE values are indicated by underlined values. The plots of test data and PAFTS forecasts are shown in Figs. 4.6, 4.7, 4.8, 4.9, 4.10, 4.11, 4.12, 4.13, 4.14 and 4.15. Table 4.3 tells that the ARIMA method produced the best MAPE value in one case (ID = 10). The MANFIS method resulted in the best RMSE value in two cases (ID = 5, 6) and the best MAPE value in one case (ID = 5). The AR–ANFIS method produced the lowest RMSE value in two cases (ID = 3, 11) and the lowest MAPE value in six cases (ID = 2, 3, 6, 7, 9,

Table 4.3 Simulation results for BIST100 time series

ID	ARIMA		ANFIS		MANFIS		AR–ANFIS		PAFTS	
	RMSE	MAPE	RMSE	MAPE	RMSE	MAPE	RMSE	MAPE	RMSE	MAPE
2	344.91	0.0087	405.15	0.0097	261	0.0059	240.39	0.0052	219.94	0.0057
3	540.21	0.0120	647.28	0.0157	503	0.0112	467.99	0.0102	472.36	0.0102
4	1221	0.0183	1141.30	0.0153	1144	0.0187	1136.70	0.0191	992.47	0.0147
5	1612	0.0219	2033	0.0277	1303	0.0193	1451.80	0.023	1473.91	0.0221
6	1057.60	0.0144	1007	0.0141	960	0.0141	987.042	0.0134	970.28	0.0143
7	1129.60	0.0150	1134	0.0153	1009	0.0136	999.532	0.013	993.60	0.0136
8	651	0.0084	634	0.0092	634	0.0082	631.94	0.0087	589.52	0.0079
9	621	0.0088	938	0.0139	629	0.0094	619.58	0.0086	609.11	0.0086
10	1361.60	0.0116	1477	0.0131	1418	0.0134	1362.3	0.0118	1347.63	0.0130
11	1268.7	0.0109	1413	0.0125	1264	0.0109	1256.46	0.0108	1263.29	0.0117

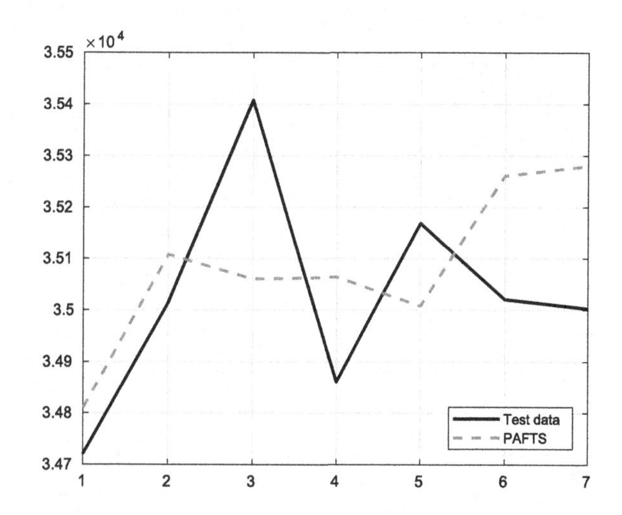

Fig. 4.6 PAFIS forecast, BIST100, 2009, ID = 2

11). As regards the PAFTS method, it gave the lowest RMSE value in six cases (ID = 2, 4, 7, 8, 9, 10) and the lowest MAPE value in four cases (ID = 3, 4, 8, 9). Based on these results, we may conclude that for the BIST100 time series the PAFTS and AR–ANFIS methods produced the best forecasts in most of the cases, and these two methods produced similarly good predictions. It is worth mentioning that the best and the second best values of both the RMSE and MAPE indicators are close to each other. The largest relative difference between the best and the second best RMSE values (14.52%) was obtained for the time series with ID = 4. As for the MAPE indicator, the largest relative difference between the best and the second best values (14.51%) was produced for the time series with ID = 5.

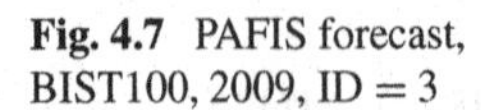
Fig. 4.7 PAFIS forecast, BIST100, 2009, ID = 3

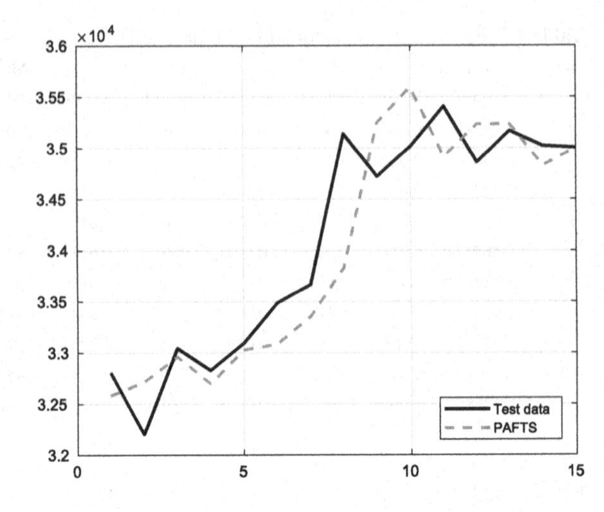

Fig. 4.8 PAFIS forecast, BIST100, 2010, ID = 4

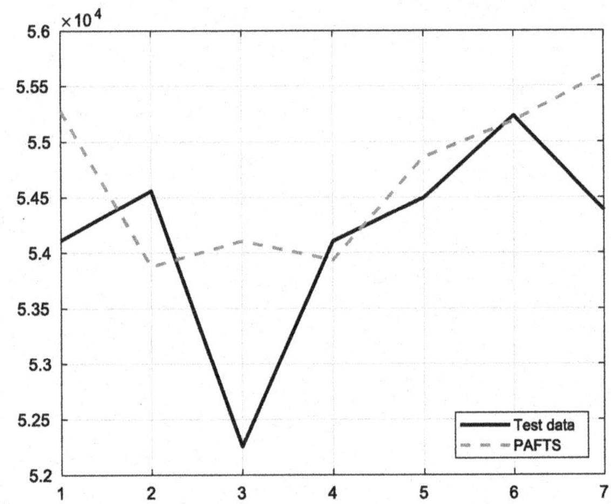

4.3.1.3 Taiwan Stock Exchange Capitalization Weighted Stock Index (TAIEX)

The last 6 of the time series in question include daily data of the Taiwan Stock Exchange Capitalization Weighted Stock Index (TAIEX) for the time period between 1999 and 2004. The properties of these time series and the parameter values of the PAFTS method applied to each of them are summarized in Table 4.4, ID = 12,...,17. The forecasting performance of the PAFTS method was compared with those of Chen [30], Chen and Chang [39], Chen and Chen [40] and Chen et al. [41]. The results of the referenced methods are from the paper of Baş et al. [7]. The PAFTS results were also compared with those of the ANFIS, MANFIS and AR–ANFIS methods. The ANFIS, MANFIS and AR–ANFIS results were taken from the paper by Sarıca et al.

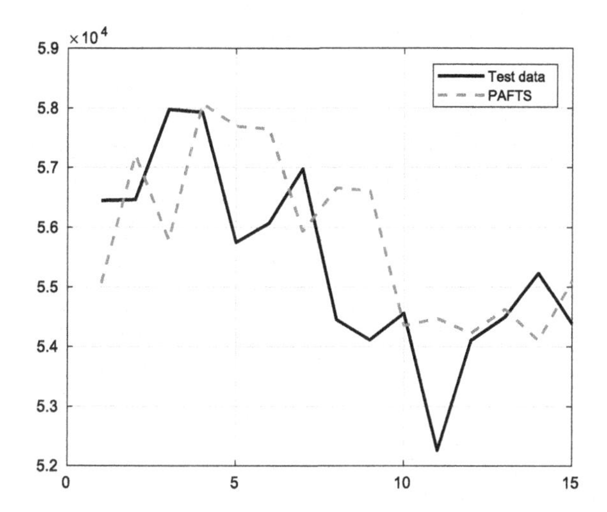

Fig. 4.9 PAFIS forecast, BIST100, 2010, ID = 5

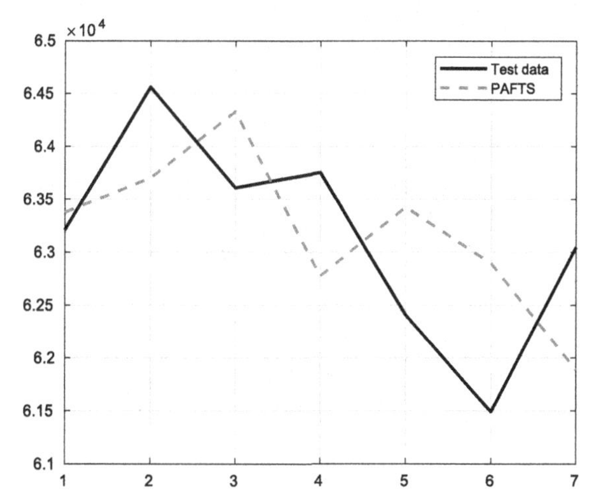

Fig. 4.10 PAFIS forecast, BIST100, 2011, ID = 6

[38]. The RMSE and MAPE values of the compared methods are shown in Table 4.4 in which the best RMSE values are indicated by underlined numbers.

It can be seen from Table 4.4 that the method of Chen et al. [41], the PAFTS, the AR–ANFIS and the MANFIS methods gave the best result in 2 (ID = 13, 17), 2 (ID = 12, 16), 1 (ID = 14) and 1 (ID = 15) cases, respectively. We also notice that the forecasting results of these four methods are close to each other. It should also be added that by taking the average of the six RMSE values of each method into account, the AR–ANFIS produced the best result with an average RMSE value of 84.00 that is just slightly better than the average RMSE value of 84.27 obtained by the PAFTS method. Based on these empirical results, it may be concluded that for the TAIEX time series the PAFTS method performs just as well as the MANFIS, AR–ANFIS and [41] methods. The plots of the test data and the PAFTS simulation

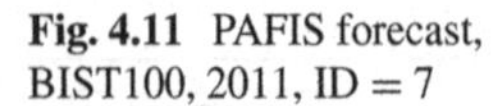

Fig. 4.11 PAFIS forecast, BIST100, 2011, ID = 7

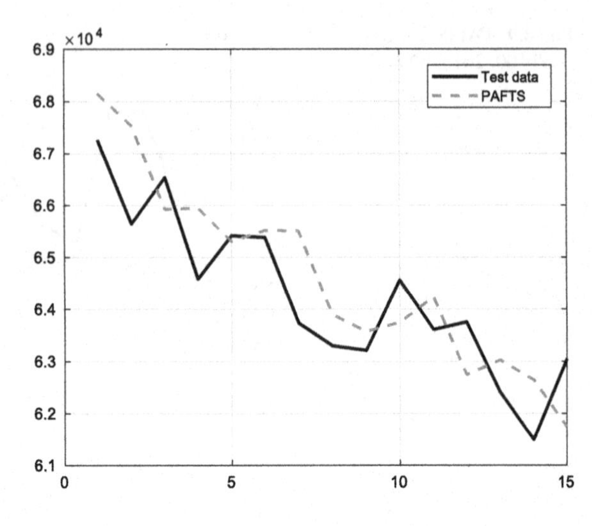

Fig. 4.12 PAFIS forecast, BIST100, 2012, ID = 8

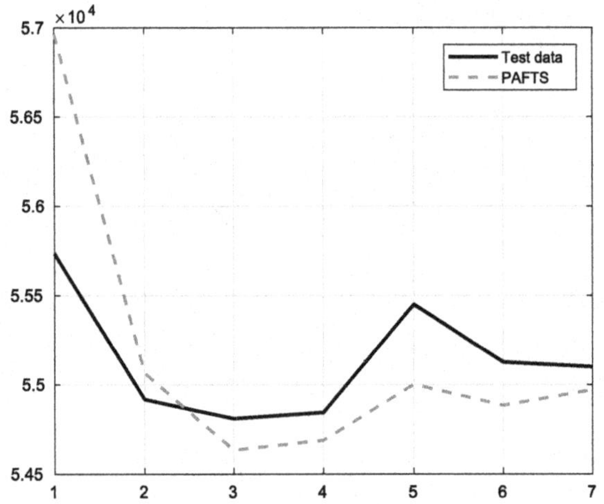

results for the TAIEX time series are shown in Figs. 4.16, 4.17, 4.18, 4.19, 4.20 and 4.21.

4.3.1.4 Summary of Forecasting Results

The forecasting capability of the PAFTS method was tested on 17 time series that are listed in Table 4.1. Based on the RMSE values, the PAFTS method was the best in eight cases (ID = 2, 4, 7, 8, 9, 10, 12, 16) out of the total 17 cases. The ANFIS, MANFIS and AR–ANFIS methods, which were also applied for all the 17 time series, resulted in the best RMSE values in 0, 3 (ID = 5, 6, 15) and 4 (ID = 1, 3, 11, 14) cases. It should also be added that in 12 cases (ID = , 2, 3, 4, 7, 8, 9, 10,

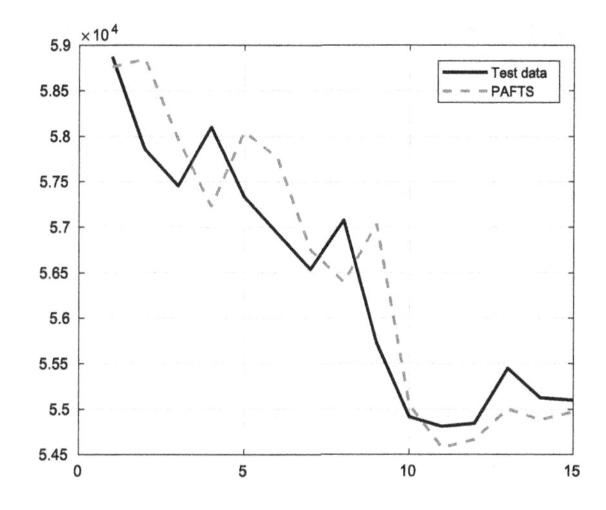

Fig. 4.13 PAFIS forecast, BIST100, 2012, ID = 9

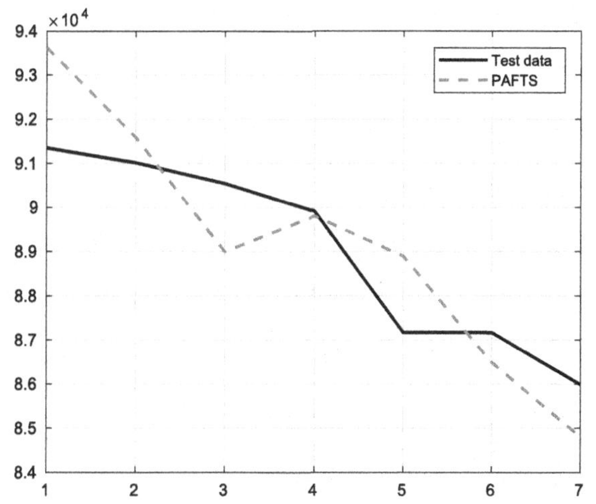

Fig. 4.14 PAFIS forecast, BIST100, 2013, ID = 10

11, 12, 14, 16), one of the AR–ANFIS and PAFTS methods produced the lowest RMSE values. Moreover, in most of these cases, where the AR–ANFIS method was the best, the PAFTS was the second best, and vice versa.

The MAPE values produced by various forecast methods were compared to those of the PAFTS method for the first 11 time series. Based on the MAPE values, the PAFTS method was the best in five cases (ID = 1, 3, 4, 8, 9), the AR–ANFIS method was the best in six cases (ID = 2, 3, 6, 7, 9, 10) and the MANFIS method was the best in one case (ID = 5). (Note that the ARIMA method was the best for ID = 10, but it was not applied for ID = 1.) It should be added that in 9 cases out of the total 11 cases (ID = 1, 2, 3, 4, 6, 7, 8, 9, 10), one of the AR–ANFIS and PAFTS methods gave the best MAPE values. We should also mention that in most of the 11 cases the AR–ANFIS and PAFTS methods produced very similar MAPE results.

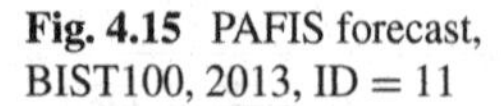

Fig. 4.15 PAFIS forecast, BIST100, 2013, ID = 11

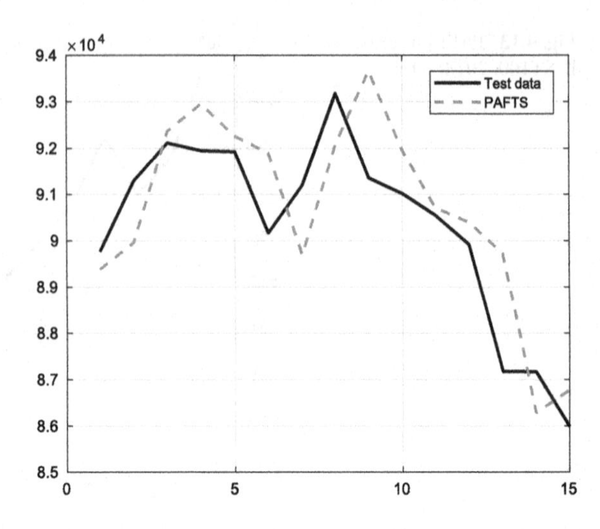

Table 4.4 RMSE values of simulations for TAIEX time series

ID	12	13	14	15	16	17	
Year Method	1999	2000	2001	2002	2003	2004	Average
[30]	120	176.32	147.84	101.18	74.46	84.28	117.34
[39]	101.97	129.42	113.33	66.82	53.51	60.48	87.58
[40]	112.47	123.62	115.33	71.01	58.06	57.73	89.70
[41]	99.87	119.98	114.47	67.17	52.49	52.27	84.37
ANFIS	101.16	137.02	114.72	65.99	57.04	61.36	89.54
MANFIS	101.94	124.92	112.47	62.57	52.33	53.66	84.64
AR–ANFIS	98.37	122.81	111.49	65.86	51.83	53.63	84.00
PAFTS	95.12	128.65	111.79	64.75	51.74	53.57	84.27

Based on the above-mentioned empirical results, the PAFTS method may be considered as a suitable alternative time series modeling technique.

4.4 Summary

In this chapter, the pliant arithmetic-based fuzzy time series model was introduced. The proposed modeling technique is founded on a fuzzy inference method in which the fuzzy output is a so-called pliant number. The novelty of our inference method lies

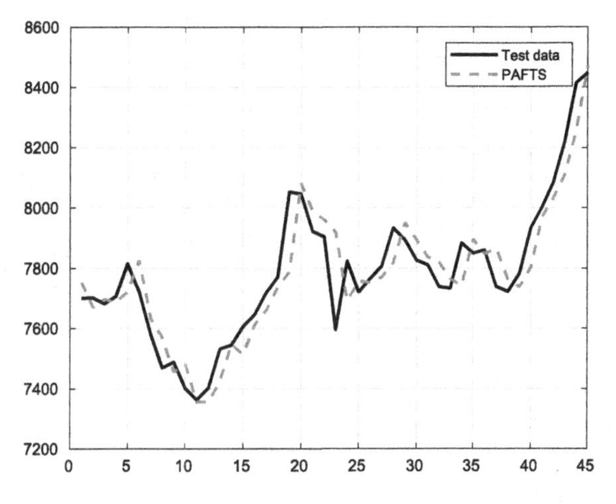

Fig. 4.16 PAFIS forecast, TAIEX, 1999, ID = 12

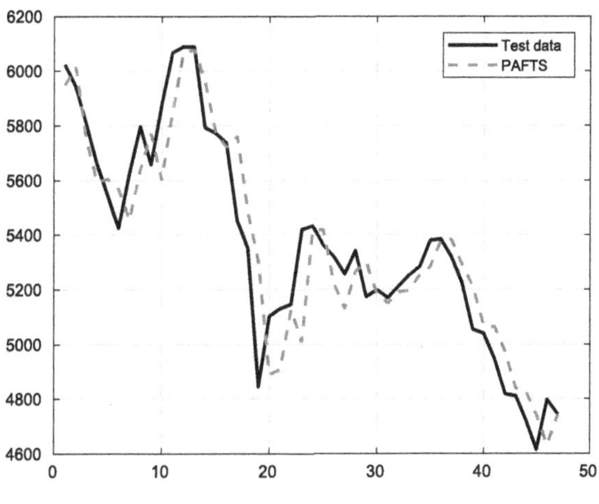

Fig. 4.17 PAFIS forecast, TAIEX, 2000, ID = 13

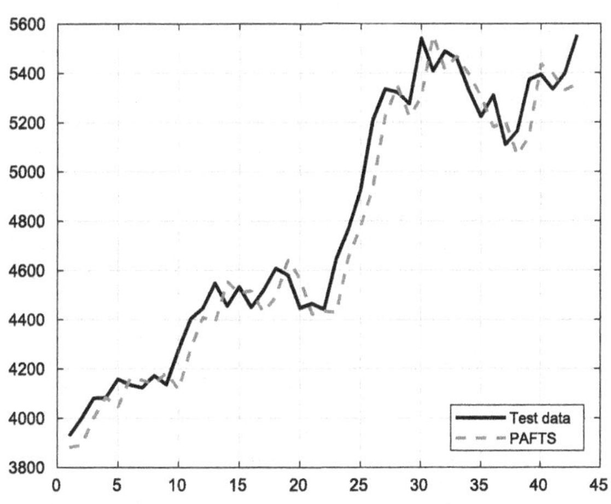

Fig. 4.18 PAFIS forecast, TAIEX, 2001, ID = 14

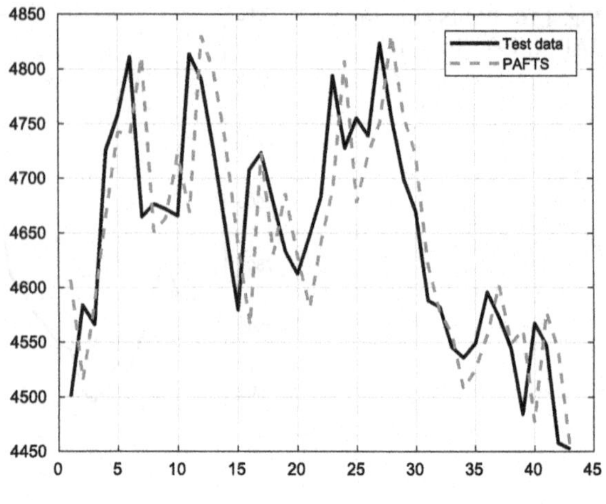

Fig. 4.19 PAFIS forecast, TAIEX, 2002, ID = 15

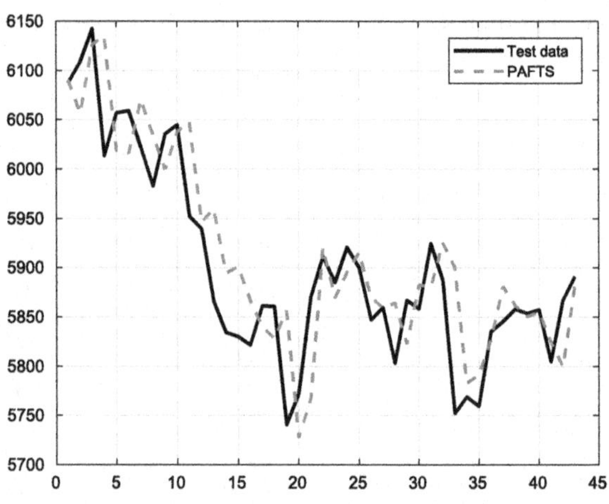

Fig. 4.20 PAFIS forecast, TAIEX, 2003, ID = 16

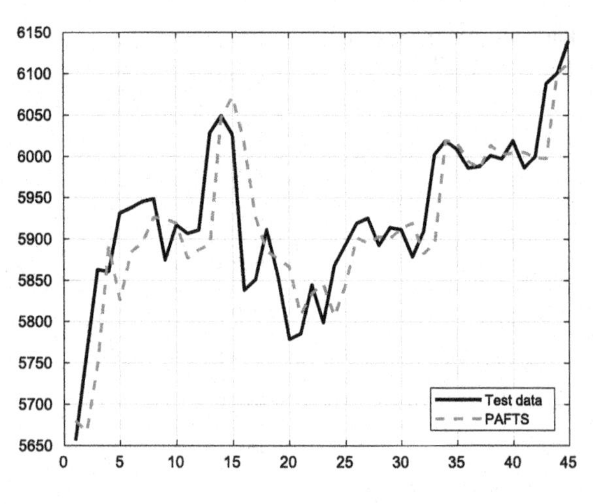

Fig. 4.21 PAFIS forecast, TAIEX, 2004, ID = 17

in the fact that its fuzzy output is obtained by fuzzy arithmetic operations; namely, via weighted summation of pliant numbers, which are the consequents of the fuzzy rules. The defuzzification method of the pliant arithmetic-based fuzzy system has two notable advantages. On the one hand, no numerical integration is needed to generate the crisp output. On the other hand, it runs in a constant time. Next, we discussed how the pliant arithmetic-based fuzzy time series (PAFTS) model can be established by utilizing the PAFIS method. The greatest advantages of our PAFTS model lies in its simplicity and easy-to-use characteristics. Once the fuzzy rule consequents are obtained, the time series modeling and forecasting become very simple. The proposed method was applied to some well-known data sets and its forecasting capability was compared with those of some familiar and recent time series modeling techniques including WMES, SARIMA, Feed Forward Artificial Neural Networks, ANFIS, MANFIS, AR–ANFIS, and with those of Chen [30], Chen and Chang [39], Chen and Chen [40] and Chen et al. [41]. Based on the empirical results, the proposed method may be viewed as novel viable time series modeling technique.

References

1. P. Singh. An efficient method for forecasting using fuzzy time series, in *Emerging Research on Applied Fuzzy Sets and Intuitionistic Fuzzy Matrices* (2016), p. 287
2. P. Singh, B. Borah, An efficient time series forecasting model based on fuzzy time series. Eng. Appl. Artif. Intell. **26**(10), 2443–2457 (2013)
3. P. Singh, A brief review of modeling approaches based on fuzzy time series. Int. J. Mach. Learn. Cybernet. **8**(2), 397–420 (2017)
4. H. Nguyen, B. Wu, *Fundamentals of Statistics with Fuzzy Data* (Springer, Berlin, 2006)
5. Q. Song, R.P. Leland, B.S. Chissom, A new fuzzy time-series model of fuzzy number observations. Fuzzy Sets Syst. **73**(3), 341–348 (1995)
6. K. Huarng, T.H.-K. Yu, The application of neural networks to forecast fuzzy time series. Phys. A **363**(2), 481–491 (2006)
7. E. Baş, E. Egrioglu, C.H. Aladag, U. Yolcu, Fuzzy-time-series network used to forecast linear and nonlinear time series. Appl. Intell. **43**(2), 343–355 (2015)
8. Q. Song, B.S. Chissom, Fuzzy time series and its models. Fuzzy Sets Syst. **54**(3), 269–277 (1993b)
9. Q. Song, B.S. Chissom, Forecasting enrollments with fuzzy time series part i. Fuzzy Sets Syst. **54**(1), 1–9 (1993a)
10. Q. Song, B.S. Chissom, Forecasting enrollments with fuzzy time series part ii. Fuzzy Sets Syst. **62**(1), 1–8 (1994)
11. K.-B. Song, Y.-S. Baek, D.H. Hong, G. Jang, Short-term load forecasting for the holidays using fuzzy linear regression method. IEEE Trans. Power Syst. **20**(1), 96–101 (2005)
12. H.J. Sadaei, R. Enayatifar, A.H. Abdullah, A. Gani, Short-term load forecasting using a hybrid model with a refined exponentially weighted fuzzy time series and an improved harmony search. Int. J. Electr. Power Energy Syst. **62**, 118–129 (2014)
13. W. Wang, W. Pedrycz, X. Liu, Time series long-term forecasting model based on information granules and fuzzy clustering. Eng. Appl. Artif. Intell. **41**, 17–24 (2015)
14. W.A. Lodwick, J. Kacprzyk, *Fuzzy Optimization: Recent Advances and Applications*, vol. 254 (Springer, Berlin, 2010)
15. L. Zadeh, From computing with numbers to computing with words-from manipulation of measurements to manipulation of perceptions. Int. J. Appl. Math. Comput. Sci. **12**(3), 307–324 (2002)

16. E. Egrioglu, C.H. Aladag, U. Yolcu, Fuzzy time series forecasting with a novel hybrid approach combining fuzzy c-means and neural networks. Expert Syst. Appl. **40**(3), 854–857 (2013)
17. U. Yolcu, C.H. Aladag, E. Egrioglu, V.R. Uslu, Time-series forecasting with a novel fuzzy time-series approach: an example for istanbul stock market. J. Stat. Comput. Simul. **83**(4), 599–612 (2013)
18. S. Sakhuja, V. Jain, S. Kumar, C. Chandra, S.K. Ghildayal, Genetic algorithm based fuzzy time series tourism demand forecast model. Ind. Manag. Data Syst. **116**(3), 483–507 (2016)
19. M.-Y. Chen, A high-order fuzzy time series forecasting model for internet stock trading. Future Gener. Comput. Syst. **37**, 461–467 (2014)
20. J.L. Salmeron, W. Froelich, Dynamic optimization of fuzzy cognitive maps for time series forecasting. Knowl.-Based Syst. **105**, 29–37 (2016)
21. K. Huarng, Effective lengths of intervals to improve forecasting in fuzzy time series. Fuzzy Sets Syst. **123**(3), 387–394 (2001)
22. S.-M. Chen, N.-Y. Chung, Forecasting enrollments of students by using fuzzy time series and genetic algorithms. Int. J. Inf. Manag. Sci. **17**(3), 1–17 (2006)
23. S.-T. Li, Y.-C. Cheng, S.-Y. Lin, A fcm-based deterministic forecasting model for fuzzy time series. Comput. Math. Appl. **56**(12), 3052–3063 (2008)
24. I.-H. Kuo, S.-J. Horng, T.-W. Kao, T.-L. Lin, C.-L. Lee, Y. Pan, An improved method for forecasting enrollments based on fuzzy time series and particle swarm optimization. Expert Syst. Appl. **36**(3), 6108–6117 (2009)
25. C.-H. Cheng, G.-W. Cheng, J.-W. Wang, Multi-attribute fuzzy time series method based on fuzzy clustering. Expert Syst. Appl. **34**(2), 1235–1242 (2008)
26. E. Egrioglu, C.H. Aladag, U. Yolcu, V.R. Uslu, M.A. Basaran, Finding an optimal interval length in high order fuzzy time series. Expert Syst. Appl. **37**(7), 5052–5055 (2010)
27. E. Egrioglu, C.H. Aladag, M.A. Basaran, U. Yolcu, V.R. Uslu, A new approach based on the optimization of the length of intervals in fuzzy time series. J. Intell. Fuzzy Syst. **22**(1), 15–19 (2011)
28. W. Lu, X. Chen, W. Pedrycz, X. Liu, J. Yang, Using interval information granules to improve forecasting in fuzzy time series. Int. J. Approx. Reason. **57**, 1–18 (2015)
29. J. Sullivan, W.H. Woodall, A comparison of fuzzy forecasting and markov modeling. Fuzzy Sets Syst. **64**(3), 279–293 (1994)
30. S.-M. Chen, Forecasting enrollments based on fuzzy time series. Fuzzy Sets Syst. **81**(3), 311–319 (1996)
31. C.H. Aladag, M.A. Basaran, E. Egrioglu, U. Yolcu, V.R. Uslu, Forecasting in high order fuzzy times series by using neural networks to define fuzzy relations. Expert Syst. Appl. **36**(3), 4228–4231 (2009)
32. E. Egrioglu, C.H. Aladag, U. Yolcu, M.A. Basaran, V.R. Uslu, A new hybrid approach based on sarima and partial high order bivariate fuzzy time series forecasting model. Expert Syst. Appl. **36**(4), 7424–7434 (2009)
33. E. Egrioglu, C.H. Aladag, U. Yolcu, V.R. Uslu, M.A. Basaran, A new approach based on artificial neural networks for high order multivariate fuzzy time series. Expert Syst. Appl. **36**(7), 10589–10594 (2009)
34. V.R. Uslu, E. Bas, U. Yolcu, E. Egrioglu, A fuzzy time series approach based on weights determined by the number of recurrences of fuzzy relations. Swarm Evol. Comput. **15**, 19–26 (2014)
35. C.H. Aladag, U. Yolcu, E. Egrioglu, A high order fuzzy time series forecasting model based on adaptive expectation and artificial neural networks. Math. Comput. Simul. **81**(4), 875–882 (2010)
36. J. Dombi, T. Jónás, Z.E. Tóth, Fuzzy time series models using pliant- and asymptotically pliant arithmetic-based inference. Neural Process. Lett. (2018). https://doi.org/10.1007/s11063-018-9927-0
37. E. Egrioglu, C. Aladag, U. Yolcu, E. Bas, A new adaptive network based fuzzy inference system for time series forecasting. Aloy J. Soft Comput. Appl. **2**, 25–32 (2014)

38. B. Sarıca, E. Egrioglu, B. Aşıkgil, A new hybrid method for time series forecasting: AR-ANFIS. Neural Comput. Appl. **29**(3), 749–760 (2018). https://doi.org/10.1007/s00521-016-2475-5
39. S.-M. Chen, Y.-C. Chang, Multi-variable fuzzy forecasting based on fuzzy clustering and fuzzy rule interpolation techniques. Inf. Sci. **180**(24), 4772–4783 (2010)
40. S.-M. Chen, C.-D. Chen, TAIEX forecasting based on fuzzy time series and fuzzy variation groups. IEEE Trans. Fuzzy Syst. **19**(1), 1–12 (2011)
41. S.-M. Chen, H.-P. Chu, T.-W. Sheu, TAIEX forecasting using fuzzy time series and automatically generated weights of multiple factors. IEEE Trans. Syst. Man Cybern.-Part A: Syst. Hum. **42**(6), 1485–1495 (2012)
42. J. Dombi, T. Jónás, and Z. E. Tóth. A pliant arithmetic-based fuzzy time series model, in *Advances in Computational Intelligence*, ed. by I. Rojas, G. Joya, A. Catala. (Springer International Publishing, Cham, 2017), pp. 129–141. https://doi.org/10.1007/978-3-319-59147-6_12
43. T. Jónás, Z. E. Tóth, and J. Dombi. Modeling failure rate time series by a fuzzy arithmetic-based inference system, in *2016 IEEE 17th International Symposium on Computational Intelligence and Informatics (CINTI)* (2016), pp. 93–98. https://doi.org/10.1109/CINTI.2016.7846385
44. T. Jónás, J. Dombi, Z.E. Tóth, P. Dömötör, Forecasting short-term demand for electronic assemblies by using soft-rules, in *Time Series Analysis and Forecasting*, ed. by I. Rojas, H. Pomares (Springer International Publishing, Cham, 2016), pp. 341–353. https://doi.org/10.1007/978-3-319-28725-6_25
45. T. Jónás, Z. Eszter Tóth, and J. Dombi. A fuzzy time series model with customized membership functions, in *Advances in Time Series Analysis and Forecasting*, ed.by I. Rojas, H. Pomares, O. Valenzuela (Springer International Publishing, Cham, 2017), pp. 285–298. https://doi.org/10.1007/978-3-319-55789-2_20
46. J. Dombi, Pliant arithmetics and pliant arithmetic operations. Acta Polytech. Hung. **6**(5), 19–49 (2009)
47. J. Dombi, Towards a general class of operators for fuzzy systems. IEEE Trans. Fuzzy Syst. **16**(2), 477–484 (2008). https://doi.org/10.1109/TFUZZ.2007.905910
48. J.C. Bezdek, *Pattern Recognition with Fuzzy Objective Function Algorithms* (Plenum Press, New York, 1981)
49. M.S. Bazaraa, H.D. Sherali, C.M. Shetty, *Nonlinear Programming: Theory and Algorithms*, 3rd edn. (Wiley, New Jersey, 2006)
50. T. Caliński, J. Harabasz, A dendrite method for cluster analysis. Commun. Stat.-Simul. Comput. **3**(1), 1–27 (1974)

Chapter 5
Likert Scale-Based Evaluations with Flexible Fuzzy Numbers

In social sciences, as well as in economics and in many other areas of science, Likert scales are generally applied for evaluating various characteristics. A Likert scale is a discrete scale, which may be viewed as an ordered finite set of pre-defined categories. In practice, the Likert scale-based evaluation results are ordinal data commonly coded by numeric values. Although the Likert scales are easy-to-use and provide the users with ordinal numeric data, their application has certain limitations. Typically, the evaluator makes complex decisions under uncertainty; however, the evaluation result is a single crisp numeric (ordinal) value. In the light of this characteristic of the Likert scale-based evaluations, some researchers have highlighted that individuals cannot appropriately express their opinion of a given situation by a single crisp value and proposed linguistic approaches to represent a specific numeric value (see, e.g. the articles [1–6]). Furthermore, when the number of values that the user can choose from on a Likert scale is not sufficiently large, the real variability associated with the rating may be lost [7]. It should be emphasized that on a Likert scale, the values (codes) are ordinal data. That is, the differences between the codes cannot always appropriately represent the differences in their magnitude. Therefore, solely statistical techniques dealing with ordinal data can be utilized to analyze the rating results, but it may lead to a loss of some relevant information [8]. Moreover, in situations where the evaluation is carried out for a given period of time, the rated item might be time dependent and so one crisp value alone cannot express the perceived variation (see, e.g. [9]). In addition, when heterogeneous preferences of the evaluators are aggregated into a single crisp score, the aggregate result tends to hide the variability associated with the rated characteristic [10].

It is well-known that fuzzy set theory and fuzzy logic can be successfully applied to situations where human perceptions, subjectivity and imprecision need to be taken into account during evaluations and decisions [11–14]. The above-mentioned shortcomings of the traditional Likert scale-based evaluations can also be mitigated by utilizing the constructions and techniques of fuzzy set theory and fuzzy logic (see, e.g. [7, 8, 15–17]). The fuzzy rating scales have the ability to model the imprecision

J. Dombi and T. Jónás, *Advances in the Theory of Probabilistic and Fuzzy Data Scientific Methods with Applications*, Studies in Computational Intelligence 814,
https://doi.org/10.1007/978-3-030-51949-0_5

and uncertainty of human evaluations through the application of fuzzy numbers (see, e.g. [6, 7, 18–21]). This approach results in fuzzy-valued responses that are able to reflect the human perceptions more precisely than a Likert scale.

In this chapter, the so-called flexible fuzzy number is introduced. The membership function of a flexible fuzzy number can exhibit a bell-shape, a triangular-shape and a 'reverse' bell-shape depending on the value of its shape parameter λ. This feature of the flexible fuzzy number allows us to express the soft equation 'x is approximately equal to x_0' in various ways. We should also point out that the 4-parameter membership function of a flexible fuzzy number may be viewed as a generalization of the triangular membership function. Here, we demonstrate that if the shape parameter λ is fixed, then the set of flexible fuzzy numbers is closed under the multiplication by scalar, fuzzy addition and weighted average operations. The pliancy of flexible fuzzy numbers and the above-mentioned properties of the operations over them make the flexible fuzzy numbers suitable for Likert scale-based fuzzy evaluations. On the one hand, we can utilize them to deal with the vagueness that may originate from the uncertainty of the evaluators or from the variability of the perceived performance values. On the other hand, in multi-dimensional Likert scale-based evaluations, by having a rating result in each evaluation dimension given by a flexible fuzzy number, these results can be easily aggregated into one flexible fuzzy number by applying the above-mentioned arithmetic operations. Furthermore, as a generalization of the flexible fuzzy numbers we introduce the extended flexible fuzzy numbers which can have different left hand side and right hand side shape parameters. We demonstrate that the asymptotic flexible fuzzy number is just a quasi fuzzy number composed of an increasing left hand side and a decreasing right hand side sigmoid function. Exploiting this property of the extended flexible fuzzy numbers allows us to perform approximate fuzzy arithmetic operations over them in the same way as in Dombi's pliant arithmetics [22]. It should be added that applications both of the flexible fuzzy numbers and the extended flexible fuzzy numbers require simple mathematical calculations and so the proposed fuzzy evaluation methods can be easily implemented and utilized in practice.

This chapter is structured as follows. In Sect. 5.1, we introduce the flexible fuzzy numbers, the extended flexible fuzzy numbers and some arithmetic operations over them, which are important from the Likert scale-based evaluation point of view. In Sect. 5.2, a demonstrative example on how the flexible fuzzy numbers can be applied to human performance evaluation is presented. Lastly, in Sect. 5.3, we summarize our main conclusions and their implications. This chapter is connected with the following publications of ours: [19, 22–24].

5.1 The Concept of Flexible Fuzzy Numbers

Here, we will use a version of the kappa function, which we introduced in Sect. 3.2.3.1, to define the membership function of the flexible fuzzy number. The triangular membership function $\mu_t(x; a, b, c)$ of a fuzzy set is commonly represented by

$$\mu_t(x;a,b,c)=\begin{cases}0, & \text{if } x \le a\\ \frac{x-a}{b-a}, & \text{if } a < x \le b\\ \frac{c-x}{c-b}, & \text{if } b < x < c\\ 0, & \text{if } c \le x,\end{cases} \tag{5.1}$$

where $a < b < c$, $x \in \mathbb{R}$. The increasing left hand side linear component of $\mu_t(x;a,b,c)$ can be written as

$$f_l(x;a,b)=\frac{x-a}{b-a}=\frac{1}{\frac{x-a+b-x}{x-a}}=\frac{1}{1+\frac{b-x}{x-a}},$$

where $x \in (a,b)$. Now, we will raise the term $\frac{b-x}{x-a}$ to the power of λ and interpret the following $\kappa(x;a,b,\lambda)$ function as a generalization of the linear function $f_l(x;a,b)$.

Definition 5.1 The kappa function $\kappa(x;a,b,\lambda)$ is given by

$$\kappa(x;a,b,\lambda)=\frac{1}{1+\left(\frac{b-x}{x-a}\right)^{\lambda}}, \tag{5.2}$$

where $a < b$, $\lambda \in \mathbb{R}$, $x \in (a,b)$.

Notice that the kappa function in Eq. (5.2) is identical with the kappa function in Eq. (3.35) if in the latter one the parameter c is set to zero. Here, we state the most important properties of the kappa function $\kappa(x;a,b,\lambda)$, namely domain, differentiability, monotonicity, limits, convexity and role of the function parameters.

Domain. The domain of the function $\kappa(x;a,b,\lambda)$ is the interval (a,b).

Differentiability. $\kappa(x;a,b,\lambda)$ is differentiable in the interval (a,b), and its derivative function is

$$\frac{\mathrm{d}\kappa(x;a,b,\lambda)}{\mathrm{d}x}=\lambda(b-a)\frac{\kappa(x;a,b,\lambda)\left(1-\kappa(x;a,b,\lambda)\right)}{(x-a)(b-x)}.$$

Monotonicity. Since $\kappa(x;a,b,\lambda) \in (0,1)$ for any $x \in (a,b)$, and using the first derivative of $\kappa(x;a,b,\lambda)$, we can state that if $\lambda > 0$, then $\kappa(x;a,b,\lambda)$ is strictly monotonously increasing, and if $\lambda < 0$, then $\kappa(x;a,b,\lambda)$ is strictly monotonously decreasing. Notice that if $\lambda = 0$, then $\kappa(x;a,b,\lambda)$ has the constant value of 0.5 for any $x \in (a,b)$.

Limits.

$$\lim_{x\to a^+}\kappa(x;a,b,\lambda)=\begin{cases}0, & \text{if } \lambda > 0\\ 1, & \text{if } \lambda < 0\end{cases}, \quad \lim_{x\to b^-}\kappa(x;a,b,\lambda)=\begin{cases}1, & \text{if } \lambda > 0\\ 0, & \text{if } \lambda < 0.\end{cases}$$

Convexity. It can be shown that the function $\kappa(x;a,b,\lambda)$ changes its shape at $x = (a+b)/2$

- from convex to concave, if $\lambda < -1$ or $\lambda > 1$
- from concave to convex, if $-1 < \lambda < 0$ or $0 < \lambda < 1$.

Notice that if $\lambda = 1$, then

$$\kappa(x; a, b, \lambda) = f_l(x; a, b) = \frac{x - a}{b - a}.$$

That is, $\kappa(x; a, b, \lambda)$ is an increasing linear function. Furthermore, if $\lambda = -1$, then $\kappa(x; a, b, \lambda)$ is the decreasing linear function

$$\kappa(x; a, b, \lambda) = f_r(x; a, b) = \frac{b - x}{b - a}.$$

Role of parameters. Parameters a and b determine the domain (a, b) of the function $\kappa(x; a, b, \lambda)$. The function $\kappa(x; a, b, \lambda)$ takes the value of 0.5 at $x = (a + b)/2$. Note that the linear functions $f_l(x; a, b)$ and $f_r(x; a, b)$ also take the value of 0.5 at $x = (a + b)/2$. The λ parameter determines the monotonicity and the shape of $\kappa(x; a, b, \lambda)$. Moreover, the slope of function $\kappa(x; a, b, \lambda)$ at $x = (a + b)/2$ is $\lambda/(b - a)$. Figure 5.1 shows some typical kappa function plots.

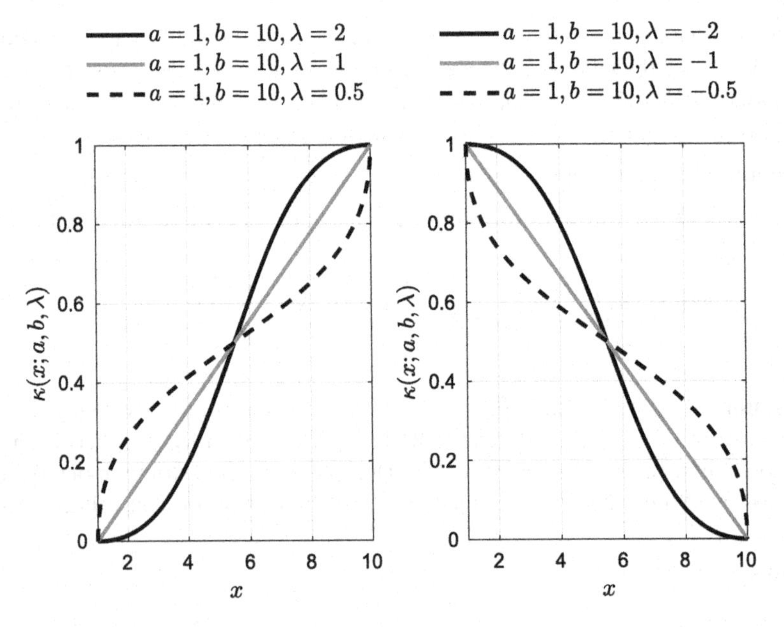

Fig. 5.1 Plots of some kappa functions

5.1.1 The Kappa Function and the Flexible Fuzzy Numbers

Utilizing the kappa function $\kappa(x; a, b, \lambda)$, we define the membership function of the so-called flexible fuzzy number as follows.

Definition 5.2 The membership function $\mu(x; l, x_0, r, \lambda)$ of the flexible fuzzy number $\{x =_{(l,r,\lambda)} x_0\}$ is given by

$$\mu(x; l, x_0, r, \lambda) = \begin{cases} 0, & \text{if } x \leq l \\ \frac{1}{1+\left(\frac{x_0-x}{x-l}\right)^{\lambda}}, & \text{if } l < x \leq x_0 \\ \frac{1}{1+\left(\frac{r-x}{x-x_0}\right)^{-\lambda}}, & \text{if } x_0 < x < r \\ 0, & \text{if } r \leq x, \end{cases}$$

where $l < x_0 < r$, $\lambda > 0$, $x \in \mathbb{R}$.

The definition of the membership function $\mu(x; l, x_0, r, \lambda)$ contains an increasing left hand side (LHS) kappa function $\kappa(x; a, b, \lambda)$ with the parameters $a = l, b = x_0$ and a decreasing right hand side (RHS) kappa function $\kappa(x; a, b, -\lambda)$, which has the parameters $a = x_0, b = r$. Notice that since $\lambda > 0$, $-\lambda$ has a negative value. In Fig. 5.2, there are some typical membership function plots of flexible fuzzy numbers. We can see from it that depending on the value of parameter λ, the membership function plot of a flexible fuzzy number can have various shapes; namely, it can be bell-shaped ($\lambda > 1$), triangular ($\lambda = 1$), or 'reverse' bell-shaped ($0 < \lambda < 1$). Hence, we call the parameter λ the shape parameter of the flexible fuzzy number. This flexibility of the membership function $\mu(x; l, x_0, r, \lambda)$ allows us to use the flexible fuzzy number $\{x =_{(l,r,\lambda)} x_0\}$ to express the soft equation 'x is approximately equal to x_0' in various ways. The membership function $\mu(x; l, x_0, r, \lambda)$ may also be viewed as a generalization of the triangular membership function. The parameters l and r determine the LHS limit and RHS limit of the flexible fuzzy number, respectively; that is, if $x \leq l$ or $x \geq r$, then the membership value of x in the fuzzy set $\{x =_{(l,r,\lambda)} x_0\}$ is zero. In other words, the truth of the statement that 'x is approximately equal to x_0' is zero, if $x \leq l$ or $x \geq r$.

The next lemma demonstrates that the membership function $\mu(x; l, x_0, r, \lambda)$ in fact defines a fuzzy number.

Lemma 5.1 *The function $\mu(x; l, x_0, r, \lambda)$ is the membership function of a fuzzy number.*

Proof The function $\mu(x; l, x_0, r, \lambda)$ is the membership function of a fuzzy number if $\mu(x; l, x_0, r, \lambda)$ has the following properties [25]:

(1) $\mu(x; l, x_0, r, \lambda)$ is normal; that is, $\exists a \in \mathbb{R}$ for which $\mu(a; l, r, \lambda) = 1$
(2) $\mu(x; l, x_0, r, \lambda)$ is fuzzy convex; that is, for any $x_1, x_2 \in \mathbb{R}$ and for any $t \in [0, 1]$, $\mu(tx_1 + (1-t)x_2; l, r, \lambda) \geq \min(\mu(x_1; l, r, \lambda), \mu(x_2; l, r, \lambda))$
(3) $\mu(x; l, x_0, r, \lambda)$ is upper semicontinuous

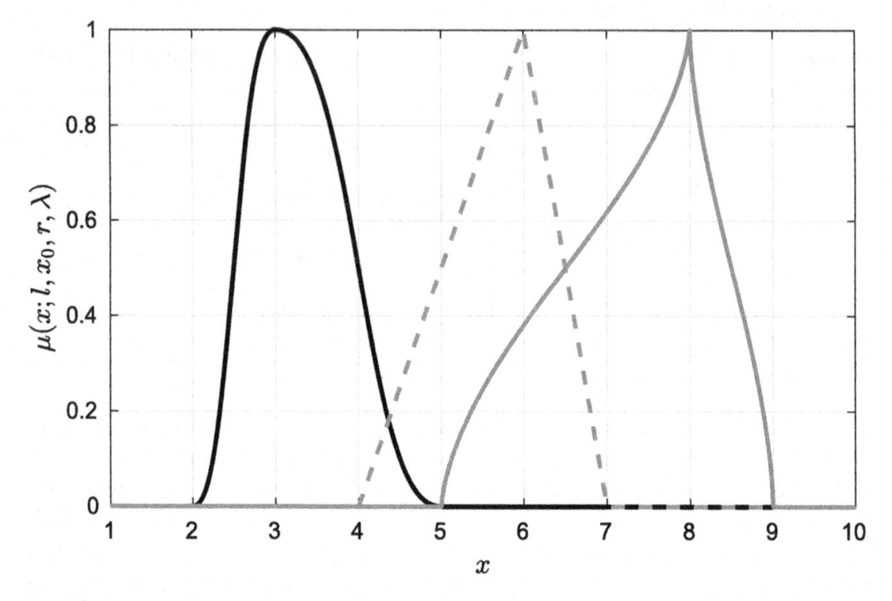

Fig. 5.2 Membership function plots of some flexible fuzzy numbers

(4) $\mu(x; l, x_0, r, \lambda)$ is compactly supported; that is, the closure of set $\{x \in \mathbb{R} : \mu(x; l, x_0, r, \lambda) > 0\}$ is compact.

Since $\mu(x_0; l, x_0, r, \lambda) = 1$ holds by definition, $\mu(x; l, x_0, r, \lambda)$ satisfies property (1). It follows from the properties of the function $\kappa(x; a, b, \lambda)$ and from the definition of $\mu(x; l, x_0, r, \lambda)$ that $\mu(x; l, x_0, r, \lambda)$ satisfies properties (2) and (3). Based on the definition of $\mu(x; l, x_0, r, \lambda)$, $\{x \in \mathbb{R} : \mu(x; l, x_0, r, \lambda) > 0\} = (l, r)$, the closure of which is compact and so $\mu(x; l, x_0, r, \lambda)$ satisfies property (4). □

We will apply the following lemma for interpreting some arithmetic operations over the set of flexible fuzzy numbers.

Lemma 5.2 *The α-cut of the flexible fuzzy number $\{x =_{(l,r,\lambda)} x_0\}$ is the interval $[x_{\alpha,l,x_0}, x_{\alpha,r,x_0}]$, where*

$$x_{\alpha,l,x_0} = \frac{x_0 + l\left(\frac{1-\alpha}{\alpha}\right)^{\frac{1}{\lambda}}}{1 + \left(\frac{1-\alpha}{\alpha}\right)^{\frac{1}{\lambda}}}, \qquad x_{\alpha,r,x_0} = \frac{x_0 + r\left(\frac{1-\alpha}{\alpha}\right)^{\frac{1}{\lambda}}}{1 + \left(\frac{1-\alpha}{\alpha}\right)^{\frac{1}{\lambda}}},$$

$l < x_0 < r$, $\lambda > 0$, $\alpha \in (0, 1)$.

Proof Based on the definition of a fuzzy α-cut (see, e.g. [26]), the α-cut of the fuzzy set $\{x =_{(l,r,\lambda)} x_0\}$ given by the membership function $\mu(x; l, x_0, r, \lambda)$ is the set $\{x \in \mathbb{R} : \mu(x; l, x_0, r, \lambda) \geq \alpha\}$. Exploiting the properties of the kappa function, it can be shown that the membership function $\mu(x; l, x_0, r, \lambda)$ takes all the real values of the interval $[0, 1]$, $\mu(x_0; l, x_0, r, \lambda) = 1$, $\mu(x; l, x_0, r, \lambda)$ is strictly increasing, if $x \leq x_0$, and it is strictly decreasing, if $x > x_0$. Furthermore, via simple calculations, we can see that $\mu(x_{l,\alpha}; l, x_0, r, \lambda) = \mu(x_{r,\alpha}; l, x_0, r, \lambda) = \alpha$. That is, the inequality $\mu(x; l, x_0, r, \lambda) \geq \alpha$ holds if and only if $x \in [x_{\alpha,l,x_0}, x_{\alpha,r,x_0}]$. This means that the α-cut of the fuzzy number $\{x =_{(l,r,\lambda)} x_0\}$ is the interval $[x_{\alpha,l,x_0}, x_{\alpha,r,x_0}]$. □

5.1.2 Some Arithmetic Operations

Next, we will describe some arithmetic operations which we will define over the set of flexible fuzzy numbers. Since from the Likert scale-based evaluation viewpoint the multiplication by a scalar, the fuzzy addition and the weighted average calculation are the most important operations, we will concentrate on interpreting these operations.

5.1.2.1 Multiplication by Scalar

Lemma 5.3 *For any $\alpha \in (0, 1)$ and $c \in \mathbb{R}$, if $\left[x_{\alpha,l,x_0}, x_{\alpha,r,x_0}\right]$ is the α-cut of the flexible fuzzy number $\{x =_{(l,r,\lambda)} x_0\}$, then $\left[cx_{\alpha,l,x_0}, cx_{\alpha,r,x_0}\right]$ is the α-cut of the flexible fuzzy number $\{x =_{(l',r',\lambda)} x_0'\}$, where $l' = cl$, $x_0' = cx_0$ and $r' = cr$.*

Proof Based on Lemma 5.2, if $\left[x_{\alpha,l,x_0}, x_{\alpha,r,x_0}\right]$ is the α-cut of the flexible fuzzy number $\{x =_{(l,r,\lambda)} x_0\}$, then

$$\left[x_{\alpha,l,x_0}, x_{\alpha,r,x_0}\right] = \left[\frac{x_0 + l\left(\frac{1-\alpha}{\alpha}\right)^{\frac{1}{\lambda}}}{1 + \left(\frac{1-\alpha}{\alpha}\right)^{\frac{1}{\lambda}}}, \frac{x_0 + r\left(\frac{1-\alpha}{\alpha}\right)^{\frac{1}{\lambda}}}{1 + \left(\frac{1-\alpha}{\alpha}\right)^{\frac{1}{\lambda}}}\right],$$

from which

$$\left[cx_{\alpha,l,x_0}, cx_{\alpha,r,x_0}\right] = \left[\frac{cx_0 + cl\left(\frac{1-\alpha}{\alpha}\right)^{\frac{1}{\lambda}}}{1 + \left(\frac{1-\alpha}{\alpha}\right)^{\frac{1}{\lambda}}}, \frac{cx_0 + cr\left(\frac{1-\alpha}{\alpha}\right)^{\frac{1}{\lambda}}}{1 + \left(\frac{1-\alpha}{\alpha}\right)^{\frac{1}{\lambda}}}\right] =$$

$$= \left[\frac{x_0' + l'\left(\frac{1-\alpha}{\alpha}\right)^{\frac{1}{\lambda}}}{1 + \left(\frac{1-\alpha}{\alpha}\right)^{\frac{1}{\lambda}}}, \frac{x_0' + r'\left(\frac{1-\alpha}{\alpha}\right)^{\frac{1}{\lambda}}}{1 + \left(\frac{1-\alpha}{\alpha}\right)^{\frac{1}{\lambda}}}\right],$$

where $l' = cl$, $x_0' = cx_0$ and $r' = cr$. Applying Lemma 5.2 again, we can see that the last interval is the α-cut of the flexible fuzzy number $\{x =_{(l',r',\lambda)} x_0'\}$. □

Lemma 5.3 can be interpreted such that for any $\alpha \in (0, 1)$, the multiplication of the α-cut of a flexible fuzzy number by a scalar is the α-cut of a flexible fuzzy number as well; that is, the set of flexible fuzzy numbers is closed under multiplication by a scalar. Exploiting this result, the following proposition holds.

Proposition 5.1 *The multiplication of the flexible fuzzy number* $\{x =_{(l,r,\lambda)} x_0\}$ *by the scalar* c *can be computed as*

$$c \odot \{x =_{(l,r,\lambda)} x_0\} = \{x =_{(l',r',\lambda)} x_0'\},$$

where $l' = cl$, $x_0' = cx_0$, $r' = cr$ *and* $\odot$ *denotes the multiplication by a scalar operator.*

Proof This proposition follows from Lemma 5.3. □

5.1.2.2 Addition

Lemma 5.4 *For any* $\alpha \in (0, 1)$, *if*

$$\left[x_{\alpha,l_1,x_{0,1}}, x_{\alpha,r_1,x_{0,1}}\right], \left[x_{\alpha,l_2,x_{0,2}}, x_{\alpha,r_2,x_{0,2}}\right], \ldots, \left[x_{\alpha,l_n,x_{0,n}}, x_{\alpha,r_n,x_{0,n}}\right]$$

are the α*-cuts of the flexible fuzzy numbers*

$$\{x =_{(l_1,r_1,\lambda)} x_{0,1}\}, \{x =_{(l_2,r_2,\lambda)} x_{0,2}\}, \ldots, \{x =_{(l_n,r_n,\lambda)} x_{0,n}\},$$

respectively, then

$$\left[\sum_{i=1}^{n} x_{\alpha,l_i,x_{0,i}}, \sum_{i=1}^{n} x_{\alpha,r_i,x_{0,i}}\right]$$

is the α*-cut of the flexible fuzzy number* $\{x =_{(l,r,\lambda)} x_0\}$, *where* $l = \sum_{i=1}^{n} l_i$, $x_0 = \sum_{i=1}^{n} x_{0,i}$ *and* $r = \sum_{i=1}^{n} r_i$.

Proof Based on Lemma 5.2, if $\left[x_{\alpha,l_i,x_{0,i}}, x_{\alpha,r_i,x_{0,i}}\right]$ is the α-cut of the flexible fuzzy number $\{x =_{(l_i,r_i,\lambda)} x_{0,i}\}$, then

$$\left[x_{\alpha,l_i,x_{0,i}}, x_{\alpha,r_i,x_{0,i}}\right] = \left[\frac{x_{0,i} + l_i\left(\frac{1-\alpha}{\alpha}\right)^{\frac{1}{\lambda}}}{1 + \left(\frac{1-\alpha}{\alpha}\right)^{\frac{1}{\lambda}}}, \frac{x_{0,i} + r_i\left(\frac{1-\alpha}{\alpha}\right)^{\frac{1}{\lambda}}}{1 + \left(\frac{1-\alpha}{\alpha}\right)^{\frac{1}{\lambda}}}\right], \tag{5.3}$$

$i = 1, 2, \ldots, n$. Summing up the equations in Eq. (5.3) for $i = 1, 2, \ldots, n$ by making use of interval arithmetic operations, we get

$$\left[\sum_{i=1}^{n} x_{\alpha,l_i,x_{0,i}}, \sum_{i=1}^{n} x_{\alpha,r_i,x_{0,i}}\right] =$$

$$= \left[\frac{\sum_{i=1}^{n} x_{0,i} + \left(\sum_{i=1}^{n} l_i\right)\left(\frac{1-\alpha}{\alpha}\right)^{\frac{1}{\lambda}}}{1+\left(\frac{1-\alpha}{\alpha}\right)^{\frac{1}{\lambda}}}, \frac{\sum_{i=1}^{n} x_{0,i} + \left(\sum_{i=1}^{n} r_i\right)\left(\frac{1-\alpha}{\alpha}\right)^{\frac{1}{\lambda}}}{1+\left(\frac{1-\alpha}{\alpha}\right)^{\frac{1}{\lambda}}}\right] =$$

$$= \left[\frac{x_0 + l\left(\frac{1-\alpha}{\alpha}\right)^{\frac{1}{\lambda}}}{1+\left(\frac{1-\alpha}{\alpha}\right)^{\frac{1}{\lambda}}}, \frac{x_0 + r\left(\frac{1-\alpha}{\alpha}\right)^{\frac{1}{\lambda}}}{1+\left(\frac{1-\alpha}{\alpha}\right)^{\frac{1}{\lambda}}}\right],$$

where $l = \sum_{i=1}^{n} l_i$, $x_0 = \sum_{i=1}^{n} x_{0,i}$ and $r = \sum_{i=1}^{n} r_i$. Next, exploiting Lemma 5.2, the last interval is the α-cut of the flexible fuzzy number $\{x =_{(l,r,\lambda)} x_0\}$. □

Lemma 5.4 tells us that for any $\alpha \in (0, 1)$, the sum of α-cuts of flexible fuzzy numbers is the α-cut of a flexible fuzzy number; that is, the set of flexible fuzzy numbers is closed under fuzzy addition. Noting this result, the following proposition can be stated on the addition of flexible fuzzy numbers.

Proposition 5.2 *The sum of the flexible fuzzy numbers*

$$\{x =_{(l_1,r_1,\lambda)} x_{0,1}\}, \{x =_{(l_2,r_2,\lambda)} x_{0,2}\}, \ldots, \{x =_{(l_n,r_n,\lambda)} x_{0,n}\},$$

can be calculated as

$$\{x =_{(l_1,r_1,\lambda)} x_{0,1}\} \oplus \{x =_{(l_2,r_2,\lambda)} x_{0,2}\} \oplus \cdots \oplus \{x =_{(l_n,r_n,\lambda)} x_{0,n}\} =$$
$$= \{x =_{(l,r,\lambda)} x_0\},$$

where $l = \sum_{i=1}^{n} l_i$, $x_0 = \sum_{i=1}^{n} x_{0,i}$, $r = \sum_{i=1}^{n} r_i$ *and* $\oplus$ *denotes the fuzzy addition operator.*

Proof This proposition follows from Lemma 5.4. □

5.1.2.3 Weighted Average

Proposition 5.3 *The weighted average*

$$w_1 \odot \{x =_{(l_1,r_1,\lambda)} x_{0,1}\} \oplus w_2 \odot \{x =_{(l_2,r_2,\lambda)} x_{0,2}\} \oplus \cdots$$
$$\cdots \oplus w_n \odot \{x =_{(l_n,r_n,\lambda)} x_{0,n}\}$$

of the flexible fuzzy numbers

$$\{x =_{(l_1,r_1,\lambda)} x_{0,1}\}, \{x =_{(l_2,r_2,\lambda)} x_{0,2}\}, \ldots, \{x =_{(l_n,r_n,\lambda)} x_{0,n}\}$$

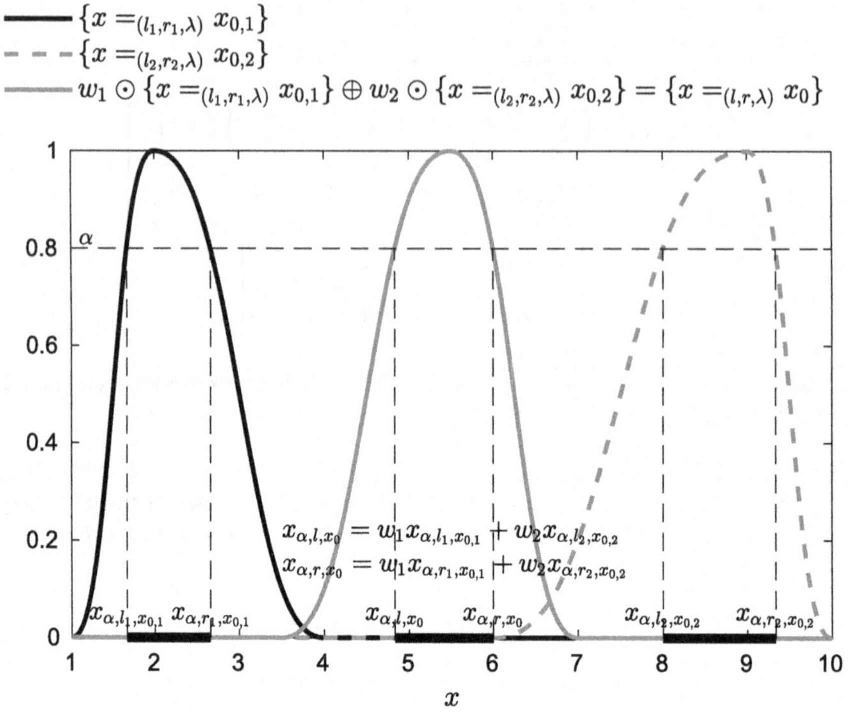

Fig. 5.3 Weighted average of two flexible fuzzy numbers

with the weights $w_1, w_2, \ldots, w_n \geq 0$, *respectively,* ($\sum_{i=1}^{n} w_i = 1$) *can be computed as*

$$w_1 \odot \{x =_{(l_1,r_1,\lambda)} x_{0,1}\} \oplus w_2 \odot \{x =_{(l_2,r_2,\lambda)} x_{0,2}\} \oplus \cdots$$
$$\cdots \oplus w_n \odot \{x =_{(l_n,r_n,\lambda)} x_{0,n}\} = \{x =_{(l,r,\lambda)} x_0\},$$

where $l = \sum_{i=1}^{n} w_i l_i$, $x_0 = \sum_{i=1}^{n} w_i x_{0,i}$ *and* $r = \sum_{i=1}^{n} w_i r_i$.

Proof This proposition directly follows from Proposition 5.1 and Proposition 5.2. □

Figure 5.3 shows plots of membership functions of the flexible fuzzy numbers $\{x =_{(l_1,r_1,\lambda)} x_{0,1}\}$ and $\{x =_{(l_2,r_2,\lambda)} x_{0,2}\}$, and a plot of the membership function of their weighted average $w_1 \odot \{x =_{(l_1,r_1,\lambda)} x_{0,1}\} \oplus w_2 \odot \{x =_{(l_2,r_2,\lambda)} x_{0,2}\}$.

Below, we will introduce some alternative notations for flexible fuzzy numbers and for their membership functions. These alternative notations may be useful in practical applications.

5.1.3 An Alternative Notation

As we pointed out before, the flexible fuzzy number $\{x =_{(l,r,\lambda)} x_0\}$ is a representation of the soft equation 'x is equal to x_0'. Thus, the membership function $\mu(x; l, x_0, r, \lambda)$ of the flexible fuzzy number $\{x =_{(l,r,\lambda)} x_0\}$ may be viewed as a function that expresses the truth of statement that 'x is equal to x_0'. The parameters l and r determine the LHS limit and RHS limit of the flexible fuzzy number $\{x =_{(l,r,\lambda)} x_0\}$, respectively; namely, the triple (l, x_0, r) determines $\{x =_{(l,r,\lambda)} x_0\}$, if the parameter λ has a given value. Notice that the properties of the above-mentioned arithmetic operations hold only if the parameter λ has a fixed value. Here, we will use the alternative notation

$$N^{(\lambda)}(l, x_0, r)$$

for the flexible fuzzy number $\{x =_{(l,r,\lambda)} x_0\}$. With this notation, the above operations can be written in the following forms.

1. Multiplication by scalar:

$$c \odot N^{(\lambda)}(l, x_0, r) = N^{(\lambda)}(cl, cx_0, cr).$$

2. Addition:

$$N^{(\lambda)}(l_1, x_{0,1}, r_1) \oplus N^{(\lambda)}(l_2, x_{0,2}, r_2) \oplus \cdots \oplus N^{(\lambda)}(l_n, x_{0,n}, r_n) = \\ = N^{(\lambda)}(l, x_0, r),$$

where $l = \sum_{i=1}^{n} l_i$ $x_0 = \sum_{i=1}^{n} x_{0,i}$ and $r = \sum_{i=1}^{n} r_i$.

3. Weighted average:

$$w_1 \odot N^{(\lambda)}(l_1, x_{0,1}, r_1) \oplus w_2 \odot N^{(\lambda)}(l_2, x_{0,2}, r_2) \oplus \cdots \\ \cdots \oplus w_n \odot N^{(\lambda)}(l_n, x_{0,n}, r_n) = N^{(\lambda)}(l, x_0, r),$$

where $l = \sum_{i=1}^{n} w_i l_i, x_0 = \sum_{i=1}^{n} w_i x_{0,i}, r = \sum_{i=1}^{n} w_i r_i, w_1, w_2, \ldots w_n \geq 0$ and $\sum_{i=1}^{n} w_i = 1$.

5.1.4 Extended Flexible Fuzzy Numbers and Approximate Operations

Previously we mentioned that the properties of the above-mentioned arithmetic operations over the set of flexible fuzzy numbers hold only if the parameter λ has a fixed value. However, there may be practical cases where arithmetic operations over flexible fuzzy numbers with various λ parameters should be performed. For example, imagine the case where the performance of a company's supplier is evaluated in multiple dimensions of the performance using flexible fuzzy numbers. Here, it may be advantageous if in different dimensions of the supplier performance the perceived performance could be represented by flexible fuzzy numbers which have different λ parameter values. Furthermore, in some situations, it would also be helpful, if the LHS and the RHS of a flexible fuzzy number had different λ parameter values. Taking these considerations into account, we shall now introduce the so-called extended flexible fuzzy numbers.

Definition 5.3 The membership function $\mu(x; l, x_0, r, \lambda_l, \lambda_r)$ of the extended flexible fuzzy number $N^{(\lambda_l,\lambda_r)}(l, x_0, r)$ is given by

$$\mu(x; l, x_0, r, \lambda_l, \lambda_r) = \begin{cases} 0, & \text{if } x \leq l \\ \frac{1}{1+\left(\frac{x_0-x}{x-l}\right)^{\lambda_l}}, & \text{if } l < x \leq x_0 \\ \frac{1}{1+\left(\frac{r-x}{x-x_0}\right)^{-\lambda_r}}, & \text{if } x_0 < x < r \\ 0, & \text{if } r \leq x, \end{cases} \tag{5.4}$$

where $l < x_0 < r$, $\lambda_l, \lambda_r > 0$, $x \in \mathbb{R}$.

The fuzzy arithmetic operations for the flexible fuzzy numbers, which we discussed in Sect. 5.1.2, exploit the fact that the λ parameter is fixed. Hence, these operations cannot be directly applied to the extended flexible fuzzy numbers. Here, we will use the sigmoid function and some of its key properties to implement approximate operations over the set of extended flexible fuzzy numbers.

Definition 5.4 The sigmoid function $\sigma_m^{(\lambda_\sigma)}(x)$ is given by

$$\sigma_m^{(\lambda_\sigma)}(x) = \frac{1}{1 + e^{-\lambda_\sigma(x-m)}}, \tag{5.5}$$

where $x \in \mathbb{R}$, $\lambda_\sigma, m \in \mathbb{R}$ and λ_σ is nonzero.

Notice that if $\lambda_\sigma = \lambda$ and $m = a$, then the sigmoid function in Eq. (5.5) coincides with the sigmoid function in Eq. (4.1). This function can neither take the value of 0, nor the value of 1, as these are its limits. However, if we choose λ_σ and m such that

$$\lambda_\sigma = \frac{2}{b-a} \ln\left(\frac{1-\varepsilon}{\varepsilon}\right), \qquad m = \frac{a+b}{2},$$

where $a < b$, $0 < \varepsilon < 0.5$ and $\varepsilon \approx 0$, e.g. $\varepsilon = 0.001$, then $\sigma_m^{(\lambda_\sigma)}(a) = \varepsilon$ and $\sigma_m^{(\lambda_\sigma)}(b) = 1 - \varepsilon$. The next proposition demonstrates an important asymptotic connection between the kappa function $\kappa(x; a, b, \lambda)$ and the sigmoid function $\sigma_m^{(\lambda_\sigma)}(x)$.

Proposition 5.4 *For any $x \in (a, b)$, if $m = (a + b)/2$ and $\lambda_\sigma = 4\lambda/(b - a)$, then*

$$\lim_{\lambda\to\infty} \left(\sigma_m^{(\lambda_\sigma)}(x) - \kappa(x; a, b, \lambda)\right) = 0. \tag{5.6}$$

Proof Utilizing the definitions and the first derivatives of the functions $\sigma_m^{(\lambda_\sigma)}(x)$ and $\kappa(x; a, b, \lambda)$, it can be shown that if $m = (a + b)/2$ and $\lambda_\sigma = 4\lambda/(b - a)$, then $\sigma_m^{(\lambda_\sigma)}(x)$ and $\kappa(x; a, b, \lambda)$ are identical to first order at $x = m$. Furthermore, since $\lambda_\sigma = 4\lambda/(b - a)$, if $\lambda \to \infty$, then $\lambda_\sigma \to \infty$ as well.

Let x have a fixed value in the interval (a, b), and let A_x and B_x be defined as follows:

$$A_x = \mathrm{e}^{-(x-m)}, \qquad B_x = \left(\frac{b-x}{x-a}\right)^{\frac{b-a}{4}}.$$

Here, we will examine three cases; namely (1) $x < m$, (2) $x = m$, (3) $x > m$.

(1) If $x < m$, then $A_x > 1$ and $B_x > 1$. As $\lambda_\sigma \to \infty$, we may exploit the fact that $\lambda_\sigma > 0$. Here, we will distinguish three sub-cases.

(a) If $A_x > B_x > 1$, then $A_x^{\lambda_\sigma} > B_x^{\lambda_\sigma} > 1$, and so

$$\lim_{\lambda_\sigma \to \infty} \left(\sigma_m^{(\lambda_\sigma)}(x) - \kappa(x; a, b, \lambda)\right) = = \lim_{\lambda_\sigma \to \infty} \left(\frac{1}{1 + A_x^{\lambda_\sigma}} - \frac{1}{1 + B_x^{\lambda_\sigma}}\right) = = \lim_{\lambda_\sigma \to \infty} \left(\frac{\left(\frac{B_x}{A_x}\right)^{\lambda_\sigma} - 1}{\left(\frac{1}{A_x^{\lambda_\sigma}} + 1\right)\left(1 + B_x^{\lambda_\sigma}\right)}\right) = 0.$$

(b) If $A_x = B_x$, then Eq. (5.6) trivially follows.

(c) If $1 < A_x < B_x$, then $1 < A_x^{\lambda_\sigma} < B_x^{\lambda_\sigma}$, and so

$$\lim_{\lambda_\sigma \to \infty} \left(\sigma_m^{(\lambda_\sigma)}(x) - \kappa(x; a, b, \lambda)\right) = = \lim_{\lambda_\sigma \to \infty} \left(\frac{1}{1 + A_x^{\lambda_\sigma}} - \frac{1}{1 + B_x^{\lambda_\sigma}}\right) = = \lim_{\lambda_\sigma \to \infty} \left(\frac{1 - \left(\frac{A_x}{B_x}\right)^{\lambda_\sigma}}{\left(1 + A_x^{\lambda_\sigma}\right)\left(\frac{1}{B_x^{\lambda_\sigma}} + 1\right)}\right) = 0.$$

(2) If $x = m$, then $\sigma_m^{(\lambda_\sigma)}(x) = \kappa(x; a, b, \lambda) = 0.5$, and so Eq. (5.6) trivially follows.

(3) If $x > m$, then $0 < A_x < 1$ and $0 < B_x < 1$, and so

$$\lim_{\lambda_\sigma \to \infty} A_x^{\lambda_\sigma} = \lim_{\lambda_\sigma \to \infty} B_x^{\lambda_\sigma} = 0.$$

Hence, Eq. (5.6) has been proven for this case as well.

□

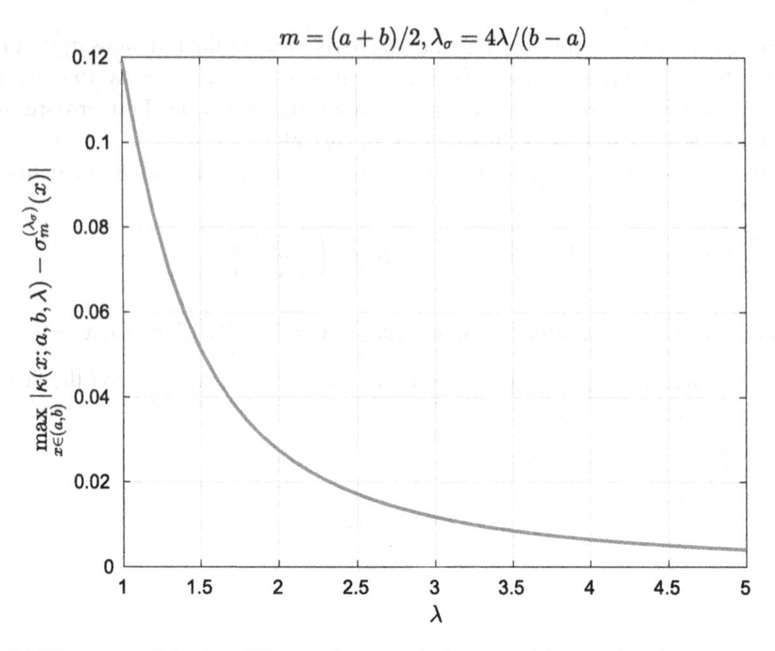

Fig. 5.4 Maximum of absolute difference between the kappa and sigmoid functions

Based on Proposition 5.4, we may conclude that if $m = (a+b)/2, \lambda_\sigma = 4\lambda/(b-a)$ and λ is large, then the kappa function $\kappa(x; a, b, \lambda)$ can be approximated well by the sigmoid function $\sigma_m^{(\lambda_\sigma)}(x)$. Figure 5.4 shows how the maximum of the absolute difference between the kappa and sigmoid functions decreases as the value of parameter λ increases.

Notice that if $\lambda > 2$, $m = (a+b)/2$ and $\lambda_\sigma = 4\lambda/(b-a)$, then

$$\max_{x\in(a,b)} \left|\kappa(x; a, b, \lambda) - \sigma_m^{(\lambda_\sigma)}(x)\right| < 0.03.$$

With the sigmoid function $\sigma_m^{(\lambda_\sigma)}(x)$, we can define the membership function of the sigmoid function-based quasi fuzzy number as follows.

Definition 5.5 The membership function $\mu_\sigma(x; m_l, x_0, m_r, \lambda_{\sigma,l}, \lambda_{\sigma,r})$ of the sigmoid function-based quasi fuzzy number $N^{(\lambda_{\sigma,l},\lambda_{\sigma,r})}(m_l, x_0, m_r)$ is given by

$$\mu_\sigma(x; m_l, x_0, m_r, \lambda_{\sigma,l}, \lambda_{\sigma,r}) = \begin{cases} \sigma_{m_l}^{(\lambda_{\sigma,l})}(x), & \text{if } x \le x_0 \\ \sigma_{m_r}^{(-\lambda_{\sigma,r})}(x), & \text{if } x_0 < x, \end{cases} \tag{5.7}$$

where $m_l < x_0 < m_r$, $\lambda_{\sigma,l}, \lambda_{\sigma,r} > 0$ and $x \in \mathbb{R}$.

Note that since the membership function $\mu_\sigma(x; m_l, x_0, m_r, \lambda_{\sigma,l}, \lambda_{\sigma,r})$ does not take the value of 1, $N^{(\lambda_{\sigma,l},\lambda_{\sigma,r})}(m_l, x_0, m_r)$ is not a fuzzy number. However, if $\lambda_{\sigma,l}$ and $\lambda_{\sigma,r}$ are sufficiently large, then $\mu_\sigma(x_0; m_l, x_0, m_r, \lambda_{\sigma,l}, \lambda_{\sigma,r}) \approx 1$.

Here, we will state an important connection between the flexible fuzzy numbers and the sigmoid function-based quasi fuzzy numbers.

Proposition 5.5 *Let* $N^{(\lambda_l,\lambda_r)}(l, x_0, r)$ *be an extended flexible fuzzy number and let* $N^{(\lambda_{\sigma,l},\lambda_{\sigma,r})}(m_l, x_0, m_r)$ *be a sigmoid function-based quasi fuzzy number. If* $m_l = (l + x_0)/2$, $m_r = (x_0 + r)/2$, $\lambda_{\sigma,l} = 4\lambda_l/(x_0 - l)$, $\lambda_{\sigma,r} = 4\lambda_r/(r - x_0)$ *and* λ_l, λ_r *are both large, then* $N^{(\lambda_l,\lambda_r)}(l, x_0, r) \approx N^{(\lambda_{\sigma,l},\lambda_{\sigma,r})}(m_l, x_0, m_r)$.

Proof This proposition follows directly from Proposition 5.4. □

Exploiting Dombi's pliant inequality model, which uses the sigmoid function to represent soft inequalities, and the properties of the pliant arithmetic operations over pliant inequalities, the following proposition can be stated.

Proposition 5.6 *The multiplication by a scalar, the addition and the weighted average operations over the set of sigmoid function-based quasi fuzzy numbers can be performed as follows.*

(1) Multiplication by a scalar:

$$c \odot N^{(\lambda_{\sigma,l},\lambda_{\sigma,r})}(m_l, x_0, m_r) = N^{(\lambda'_{\sigma,l},\lambda'_{\sigma,r})}(m'_l, x'_0, m'_r),$$

where $\lambda'_{\sigma,l} = \lambda_{\sigma,l}/c$, $\lambda'_{\sigma,r} = \lambda_{\sigma,r}/c$, $m'_l = cm_l$, $x'_0 = cx_0$, $m'_r = cm_r$ *and* $\odot$ *denotes the multiplication by a scalar operator.*

(2) Addition:

$$N^{(\lambda_{\sigma,l,1},\lambda_{\sigma,r,1})}(m_{l,1}, x_{0,1}, m_{r,1}) \oplus N^{(\lambda_{\sigma,l,2},\lambda_{\sigma,r,2})}(m_{l,2}, x_{0,2}, m_{r,2}) \oplus \cdots$$
$$\cdots \oplus N^{(\lambda_{\sigma,l,n},\lambda_{\sigma,r,n})}(m_{l,n}, x_{0,n}, m_{r,n}) = N^{(\lambda_{\sigma,l},\lambda_{\sigma,r})}(m_l, x_0, m_r),$$

where

$$\frac{1}{\lambda_{\sigma,l}} = \sum_{i=1}^{n} \frac{1}{\lambda_{\sigma,l,i}}, \quad \frac{1}{\lambda_{\sigma,r}} = \sum_{i=1}^{n} \frac{1}{\lambda_{\sigma,r,i}},$$

$m_l = \sum_{i=1}^{n} m_{l,i}$, $x_0 = \sum_{i=1}^{n} x_{0,i}$, $m_r = \sum_{i=1}^{n} m_{r,i}$ *and* $\oplus$ *denotes the fuzzy addition operator.*

(3) Weighted average:

$$w_1 \odot N^{(\lambda_{\sigma,l,1},\lambda_{\sigma,r,1})}(m_{l,1}, x_{0,1}, m_{r,1}) \oplus$$
$$\oplus w_2 \odot N^{(\lambda_{\sigma,l,2},\lambda_{\sigma,r,2})}(m_{l,2}, x_{0,2}, m_{r,2}) \oplus \cdots$$
$$\cdots \oplus w_n \odot N^{(\lambda_{\sigma,l,n},\lambda_{\sigma,r,n})}(m_{l,n}, x_{0,n}, m_{r,n}) = N^{(\lambda_{\sigma,l},\lambda_{\sigma,r})}(m_l, x_0, m_r),$$

where

$$\frac{1}{\lambda_{\sigma,l}} = \sum_{i=1}^{n} \frac{w_i}{\lambda_{\sigma,l,i}}, \qquad \frac{1}{\lambda_{\sigma,r}} = \sum_{i=1}^{n} \frac{w_i}{\lambda_{\sigma,r,i}},$$

$m_l = \sum_{i=1}^{n} w_i m_{l,i}$, $x_0 = \sum_{i=1}^{n} w_i x_{0,i}$, $m_r = \sum_{i=1}^{n} w_i m_{r,i}$, $w_1, w_2, \ldots w_n \geq 0$ *and* $\sum_{i=1}^{n} w_i = 1$.

Proof This proposition follows from the properties of the pliant arithmetic operations over pliant inequalities (see [22]). □

Based on Proposition 5.5, we may conclude that an extended flexible fuzzy number can be approximated well by an appropriate sigmoid function-based quasi fuzzy number. Moreover, in Proposition 5.6, we demonstrated how the multiplication by scalar, addition and weighted average operations can be computed over the sigmoid function-based quasi fuzzy numbers. Exploiting these findings, the next proposition about the approximations of the examined operations over the set of extended flexible fuzzy numbers can be stated.

Proposition 5.7 *The multiplication by a scalar, the addition and the weighted average operations over the set of extended flexible fuzzy numbers can be approximated as follows.*

(1) Multiplication by a scalar: if λ_l *and* λ_r *are sufficiently large, then*

$$c \odot N^{(\lambda_l,\lambda_r)}(l, x_0, r) \approx N^{(\lambda'_l,\lambda'_r)}(l', x'_0, r'), \tag{5.8}$$

where $\lambda'_l = \lambda_l$, $\lambda'_r = \lambda_r$, $l' = cl$, $x'_0 = cx_0$, $r' = cr$ *and* $\odot$ *denotes the multiplication by a scalar operator.*

(2) Addition: if $\lambda_{l,1}, \lambda_{l,2}, \ldots, \lambda_{l,n}$ *and* $\lambda_{r,1}, \lambda_{r,2}, \ldots, \lambda_{r,n}$ *are sufficiently large, then*

$$\begin{aligned} &N^{(\lambda_{l,1},\lambda_{r,1})}(l_1, x_{0,1}, r_1) \oplus N^{(\lambda_{l,2},\lambda_{r,2})}(l_2, x_{0,2}, r_2) \oplus \cdots \\ &\cdots \oplus N^{(\lambda_{l,n},\lambda_{r,n})}(l_n, x_{0,n}, r_n) \approx N^{(\lambda_l,\lambda_r)}(l, x_0, r), \end{aligned} \tag{5.9}$$

where

$$\lambda_l = \frac{x_0 - l}{\sum\limits_{i=1}^{n} \frac{x_{0,i} - l_i}{\lambda_{l,i}}}, \qquad \lambda_r = \frac{r - x_0}{\sum\limits_{i=1}^{n} \frac{r_i - x_{0,i}}{\lambda_{r,i}}}, \tag{5.10}$$

$l = \sum_{i=1}^{n} l_i$, $x_0 = \sum_{i=1}^{n} x_{0,i}$, $r = \sum_{i=1}^{n} r_i$ *and* $\oplus$ *denotes the fuzzy addition operator.*

(3) Weighted average: if $\lambda_{l,1}, \lambda_{l,2}, \ldots, \lambda_{l,n}$ *and* $\lambda_{r,1}, \lambda_{r,2}, \ldots, \lambda_{r,n}$ *are sufficiently large, then*

$$\begin{aligned} &w_1 \odot N^{(\lambda_{l,1},\lambda_{r,1})}(l_1, x_{0,1}, r_1) \oplus w_2 \odot N^{(\lambda_{l,2},\lambda_{r,2})}(l_2, x_{0,2}, r_2) \oplus \cdots \\ &\cdots \oplus w_n \odot N^{(\lambda_{l,n},\lambda_{r,n})}(l_n, x_{0,n}, r_n) \approx N^{(\lambda_l,\lambda_r)}(l, x_0, r), \end{aligned} \tag{5.11}$$

where

$$\lambda_l = \frac{x_0 - l}{\sum_{i=1}^{n} w_i \frac{x_{0,i} - l_i}{\lambda_{l,i}}}, \qquad \lambda_r = \frac{r - x_0}{\sum_{i=1}^{n} w_i \frac{r_i - x_{0,i}}{\lambda_{r,i}}}, \tag{5.12}$$

$l = \sum_{i=1}^{n} w_i l_i, x_0 = \sum_{i=1}^{n} w_i x_{0,i}, r = \sum_{i=1}^{n} w_i r_i, w_1, w_2, \ldots w_n \geq 0 \, and \sum_{i=1}^{n} w_i = 1.$

Proof The proposition follows from Propositions 5.5 and 5.6. □

Remark 5.1 If we replace the parameters λ_l and λ_r in Eq. (5.8) by a fixed λ, then $\lambda_l' = \lambda_r' = \lambda$. Similarly, if we substitute the parameters $\lambda_{l,i}$ and $\lambda_{r,i}$ in Eq. (5.9) and in Eq. (5.11) with a fixed λ, then the right hand sides of the formulas for λ_l and λ_r in Eq. (5.10) and in Eq. (5.12) also become identical with λ, where $i = 1, 2, \ldots, n$. At the same time, the exact operations for the flexible fuzzy numbers can be utilized instead of the approximate operations for the extended flexible fuzzy numbers.

5.2 A Demonstrative Example

Now, we will demonstrate the application of flexible fuzzy numbers by means of an example. In this example, the performance of an employee of a company is evaluated on Likert scales using flexible fuzzy numbers. The company evaluates the human performance in four dimensions; namely, the quality of work, the efficiency of work, the collaboration with other colleagues and the communication with other stakeholders are evaluated by the employee's supervisor. In each performance dimension, the evaluation is carried out on a Likert scale that includes the integers from 1 through 10. Number 1 represents the worst performance, while number 10 represents the best performance. The evaluator selects the values of l, x_0 and r on the Likert scale such that in the evaluator's judgement the employee's performance (1) is not worse than l, (2) mostly has the value of x_0, (3) is not better than r. These three numbers and a pre-set value of the parameter λ determine a flexible fuzzy number.

Figure 5.5 shows an evaluation example in which the quality, the efficiency, the collaboration and the communication of an employee is represented by the flexible fuzzy numbers $N^{(\lambda)}(2, 5, 6)$, $N^{(\lambda)}(5, 8, 9)$, $N^{(\lambda)}(3, 4, 5)$ and $N^{(\lambda)}(1, 3, 5)$, respectively, while the parameter λ has the value of 2. The flexible fuzzy number $N^{(\lambda)}(l, x_0, r)$ represents the truth value of the statement that 'x is equal to x_0'. This truth value is 0 if $x \leq l$ or $x \geq r$, and it is 1 if $x = x_0$. Therefore, for example, the fuzzy number $N^{(\lambda)}(2, 5, 6)$, which represents the performance in the work quality dimension, expresses the fact that in the evaluator's opinion the employee's performance is around 5, but it is not worse than 2, and it is not better than 6. It should be mentioned that the flexible fuzzy number-based evaluation carries much more information than the traditional evaluation which would represent the employee's work quality by the crisp value of 5. Here, the flexible fuzzy number-based evaluation expresses the fact that the perceived performance is between 2 and 6, and it usually has the value of 5. It

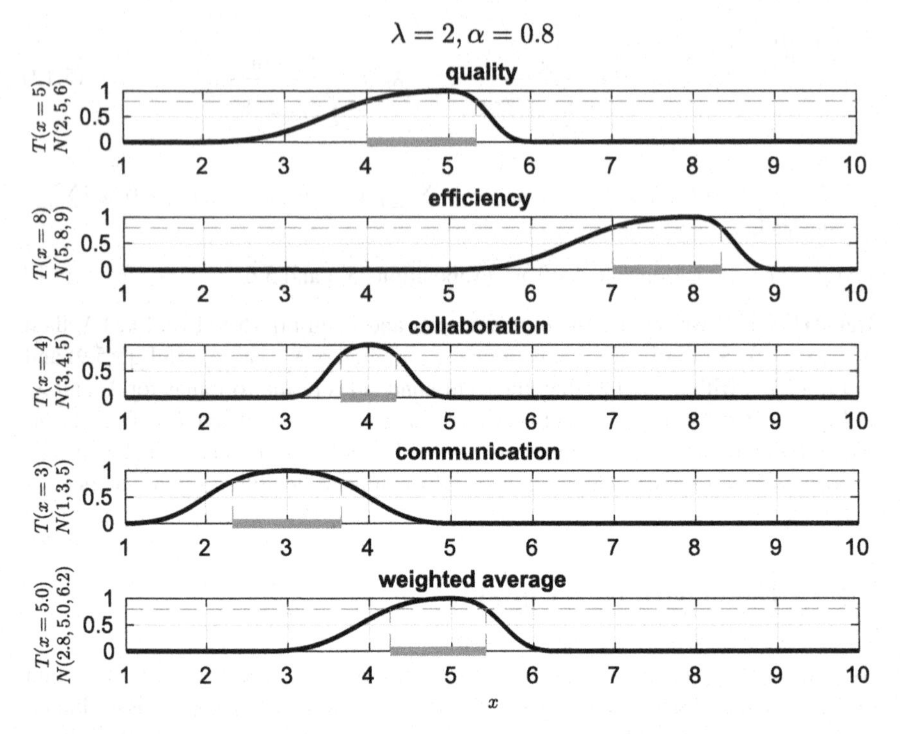

Fig. 5.5 An example of Likert Scale-based evaluations with flexible fuzzy numbers

should also be added that since the left hand limit of the fuzzy number $N^{(\lambda)}(2, 5, 6)$ is more distant from 5 than its right hand limit, the employee's performance is more likely to be less than 5 than it is greater than 5. In general, the $r - l$ width of the flexible fuzzy number $N^{(\lambda)}(l, x_0, r)$ may be viewed as an indication of performance instability, while the asymmetry of $N^{(\lambda)}(l, x_0, r)$ provides information about the direction that the performance more likely tends to differ from the value of x_0. It means that utilizing a flexible fuzzy number for evaluation on a Likert scale allows the evaluator to express his or her uncertainty about the performance quite well.

Having determined the flexible fuzzy number-based evaluation for each performance dimension, the employee's aggregate performance is computed as the weighted average of the dimension specific evaluation results. That is, the weighted average of four flexible fuzzy numbers gives the aggregate performance results. Here, each performance dimension was taken with the same weight and so the aggregate result was computed as the arithmetic average of the dimension specific evaluation results. The aggregate performance is shown in the lowest plot of Fig. 5.5. The flexible fuzzy number $N^{(\lambda)}(2.8, 5.0, 6.2)$, which represents the aggregate performance of the evaluated employee, can be interpreted as the aggregate performance of the employee lying between 2.8 and 6.2, and its most likely value is 5.

The grey colored horizontal line segment in each plot of Fig. 5.5 represents the α cut of the corresponding flexible fuzzy number for $\alpha = 0.8$. In the case of the first performance dimension, the α cut of the flexible fuzzy number $N^{(\lambda)}(2, 5, 6)$ is the interval $[4, 5.3]$. It can be interpreted such that if the performance is in the interval $[4, 5.3]$, then the truth of the statement that the perceived performance equals 5 is at least 0.8. Recall that we proved earlier that the weighted average of α cuts of flexible fuzzy numbers is the α cut of the aggregate flexible fuzzy number. Note that the values of parameters l, x_0 and r of the aggregate fuzzy number are independent of the value of parameter λ, but the α cuts of the individual flexible fuzzy numbers and the α cut of the aggregate flexible fuzzy number all depend on the value of λ.

It is worth mentioning that if the company carries out a performance evaluation for all its employees by applying the above-described approach, then the company level aggregate performance can be computed as the average of the flexible fuzzy numbers representing the aggregate performance of the individuals. Since the weighted average calculation over the set of flexible fuzzy numbers is very simple (see Proposition 5.3), it is easy to determine the company level aggregate performance.

It should be added that all the above-mentioned performance evaluations could also be carried out by applying extended flexible fuzzy numbers instead of flexible fuzzy numbers. Note that in this case, the approximate operators discussed in Proposition 5.7 should be used to perform calculations over the extended flexible fuzzy numbers.

5.3 Summary

In this chapter, we introduced the flexible fuzzy number, the membership function of which can have various shapes depending on its shape parameter. We pointed out that this feature of the flexible fuzzy number makes it suitable for expressing the soft equation 'x is approximately equal to x_0' in various ways. We showed that the triangular membership function is just a special case of the membership function of the flexible fuzzy number. Next, we proved that if the shape parameter λ is fixed, then the set of flexible fuzzy numbers is closed under the multiplication by scalar, fuzzy addition and weighted average operations. Moreover, as a generalization of the flexible fuzzy numbers we introduced the extended flexible fuzzy numbers which can have different LHS and RHS shape parameters. Here, we pointed out that the asymptotic flexible fuzzy number is just a quasi fuzzy number composed of an increasing left hand side and a decreasing right hand side sigmoid function. Exploiting this finding and Dombi's pliant arithmetic operations [22], we implemented easy-to-compute approximate fuzzy arithmetic operations over the extended flexible fuzzy numbers. By means of a human performance evaluation example, we demonstrated that the flexible fuzzy numbers can handle the vagueness that may originate from the uncertainty of the evaluators or from the variability of the perceived performance values. This pliancy of the flexible fuzzy numbers and the above-mentioned properties of the operations over them make these fuzzy numbers suitable for Likert scale-based fuzzy evaluations. Since multiple flexible fuzzy numbers can easily be aggregated into one flexible fuzzy number by performing a weighted average operation over them, the flexible fuzzy numbers can be readily employed in multi-dimensional Likert scaled-based evaluations.

References

1. F. Herrera, E. Herrera-Viedma, Choice functions and mechanisms for linguistic preference relations. Eur. J. Oper. Res. **120**(1), 144–161 (2000)
2. F. Herrera, E. López, C. Mendana, M.A. Rodríguez, Solving an assignment-selection problem with verbal information and using genetic algorithms1. Eur. J. Oper. Res. **119**(2), 326–337 (1999)
3. J. Kacprzyk, Towards "human-consistent" multistage decision making and control models using fuzzy sets and fuzzy logic. Fuzzy Sets Syst. **18**(3), 299–314 (1986)
4. S. Andayani, S. Hartati, R. Wardoyo, D. Mardapi, Decision-making model for student assessment by unifying numerical and linguistic data. Int. J. Electr. Comput. Eng. **7**(1), 363 (2017)
5. R.A. Carrasco, P. Villar, M.J. Hornos, E. Herrera-Viedma, A linguistic multi-criteria decision making model applied to the integration of education questionnaires. Int. J. Comput. Intell. Syst. **4**(5), 946–959 (2011)
6. C.T. Chen, Applying linguistic decision-making method to deal with service quality evaluation problems. Int. J. Uncertain. Fuzziness Knowl.-Based Syst. **9**(supp01), 103–114 (2001)
7. M.Á. Gil, G. González-Rodríguez, Fuzzy vs. Likert scale in statistics, in *Combining Experimentation and Theory* (Springer, 2012), pp. 407–420
8. M.A. Lubiano, S.d. l.R. de Sáa, M. Montenegro, B. Sinova, M. Á. Gil. Descriptive analysis of responses to items in questionnaires. Why not using a fuzzy rating scale? Inf. Sci. **360**, 131–148 (2016)
9. Z.E. Tóth, V. Surman, G. Árva, Challenges in course evaluations at Budapest University of Technology and Economics, in *8th ICEEPSY 2017 The International Conference on Education and Educational Psychology*, ed. by Z. Bekirogullari, M.Y. Minas, R.X. Thambusamy (2017), pp. 629–641. https://doi.org/10.15405/epsbs.2017.10.60
10. M. Kuzmanovic, G. Savic, M. Popovic, M. Martic, A new approach to evaluation of university teaching considering heterogeneity of students' preferences. High. Educ. **66**(2), 153–171 (2013)
11. H.-T. Lin, Fuzzy application in service quality analysis: an empirical study. Expert Syst. Appl. **37**(1), 517–526 (2010)
12. W.-J. Deng, Fuzzy importance-performance analysis for determining critical service attributes. Int. J. Serv. Ind. Manag. **19**(2), 252–270 (2008)
13. T. Lupo, A fuzzy servqual based method for reliable measurements of education quality in Italian higher education area. Expert Syst. Appl. **40**(17), 7096–7110 (2013)
14. T. Lupo, A fuzzy framework to evaluate service quality in the healthcare industry: an empirical case of public hospital service evaluation in Sicily. Appl. Soft Comput. **40**, 468–478 (2016)
15. Q. Li, A novel Likert scale based on fuzzy sets theory. Expert Syst. Appl. **40**(5), 1609–1618 (2013)
16. P. Quirós, J.M. Alonso, D.P. Pancho, Descriptive and comparative analysis of human perceptions expressed through fuzzy rating scale-based questionnaires. Int. J. Comput. Intell. Syst. **9**(3), 450–467 (2016)
17. B. Hesketh, R. Pryor, M. Gleitzman, T. Hesketh, Practical applications and psychometric evaluation of a computerised fuzzy graphic rating scale, in *Advances in Psychology*, vol. 56 (Elsevier, 1988), pp. 425–454
18. M.Á. Gil, M.A. Lubiano, S.D.L.R. De Sáa, B. Sinova, Analyzing data from a fuzzy rating scale-based questionnaire: a case study. Psicothema **27**(2), 182–192 (2015)

19. T. Jónás, Z.E. Tóth, G. Árva, Applying a fuzzy questionnaire in a peer review process. Total Q. Manag. Bus. Excell. **29**(9–10), 1228–1245 (2018). https://doi.org/10.1080/14783363.2018.1487616
20. J. Stoklasa, T. Talášek, and P. Luukka. Fuzzified likert scales in group multiple-criteria evaluation, in *Soft Computing Applications for Group Decision-making and Consensus Modeling*, ed. by M. Collan, J. Kacprzyk, (Springer International Publishing, Cham, 2018), pp. 165–185. https://doi.org/10.1007978-3-319-60207-3_11
21. T. Talásek, J. Stoklasa, A numerical investigation of the performance of distance and similarity measures in linguistic approximation under different linguistic scales. Multiple-Valued Logic Soft Comput. **29**, 485–503 (2017)
22. J. Dombi, Pliant arithmetics and pliant arithmetic operations. Acta Polytech. Hung. **6**(5), 19–49 (2009)
23. J. Dombi, T. Jónás, Flexible fuzzy numbers for likert scale-based evaluations, in *8th International Workshop on Soft Computing Applications (SOFA)*, Arad, Romania (2018). Accessed 13–15 Sept 2018
24. J. Dombi, On a certain type of unary operators, in *2012 IEEE International Conference on Fuzzy Systems* (2012), pp. 1–7. https://doi.org/10.1109/FUZZ-IEEE.2012.6251349
25. B. Bede, *Fuzzy Numbers* (Springer, Berlin, 2013), pp. 51–64. ISBN 978-3-642-35221-8. https://doi.org/10.1007/978-3-642-35221-8_4
26. H. Zimmermann, *Fuzzy Set Theory and Its Applications* (Springer, Bücher., 2013)

Printed in the United States
by Baker & Taylor Publisher Services